Not Just in Time

Not Just in Time

The Story of Kronos Incorporated, from Concept to Global Entity

As told by Founder Mark Ain

Written by Jim Floyd

MELIORA PRESS

An imprint of the University of Rochester Press

First published 2022

Meliora Press is an imprint of the
University of Rochester Press
668 Mt. Hope Avenue, Rochester, NY 14620, USA
www.urpress.com
and Boydell & Brewer Limited
PO Box 9, Woodbridge, Suffolk IP12 3DF, UK
www.boydellandbrewer.com

ISBN-13: 978-1-64825-037-8

Cataloging-in-Publication data available from the Library of Congress

This publication is printed on acid-free paper.

Printed in the United States of America.

Dedicated

to my grandmother,

Dora Ponemon

Because success is impossible without inspiration

Love,

Your Grandson Mark

CONTENTS

FOREWORD

Mark Ain comes from a distinguished line of pioneers and entrepreneurs. His mother, Pearl, was one of the first women to graduate from Columbia Law School—several years before the late Chief
Justice Ruth Bader Ginsberg. Earlier, Pearl's mother, Dora Ponemon,
at the age of 16, told Mark's maternal grandfather, Harry, that their
betrothal hinged on his willingness to control his own destiny by
becoming his own boss. In words Dora relayed to young Mark time and
again as she mentored her bright young grandson, she told Harry she
wanted to be married to a horse and not a mule. Harry heeded Dora's
mandate, they married, and together became entrepreneurs, starting
and running several successful business ventures rather than have others boss them around.

This engaging book relays how Mark Ain followed his grandmother's advice and forsook a plodding, mulish existence in the corporate
world for the riskier and ultimately much more rewarding life of an
entrepreneurial thoroughbred. Mark's role in the creation and growth
of Kronos, a human capital management firm that now generates
over $3 billion in annual revenue and, as the Ultimate Kronos Group
(UKG), has over 12,000 employees, provides many valuable lessons.
For those seeking to either follow in Mark's entrepreneurial footsteps
or foster an entrepreneurial culture within more established organizations, the principles behind his success are instructive.

What gets less attention in the book is Mark's commitment to giving back and paying forward to others via the prosperity he has realized through his entrepreneurial leadership of Kronos. As the Vice
Provost of Entrepreneurship at the University of Rochester and as the
former dean of Mark's alma mater the Simon Business School, respec-
tively, the two of us can attest to the profound difference that Mark has
made in developing a next generation of entrepreneurial thorough-
breds—through his involvement, insight, investment, and inspiration.

We were there in the Eastman Theater audience in 2005 when
Mark gave his thoughtful and uplifting speech (reprinted in Chapter 22) to several hundred Simon students about to officially receive
their MBA degrees. Over the years that followed, we had the honor

and pleasure of working with Mark as he joined relevant advisory boards (Simon's Executive Advisory Committee and the University of Rochester's Board of Trustees); made significant philanthropic invest-ments in the University's Entrepreneurship Program; served regu-larly as a business plans competition judge; and mentored numerous entrepreneurially inclined students as well as the leaders of nascent Rochester-area ventures.

Mark creates a bond with the students that he sponsors through scholarships and with those who participate in the annual Ain Business Plan contest. He treats them as part of his extended family—attending their weddings, visiting their newly born children, investing in their companies, and continuing to mentor them long after they have left the University of Rochester. This side of Mark is not well known but is part of his DNA.

From a fairly run-of-the-mill program, the Ain Center for Entre-preneurship at the University of Rochester has been growing over the last 15 years into one of the most highly rated programs of its kind. For example, in 2018, the *Financial Times* rated the University of Rochester's Simon Business School, with which the Ain Center is affili-ated, as being in the top nine in the nation, and 19th in the world, for entrepreneurship.

Mark's enviable track record as a thoroughbred at Kronos, cou-pled with his demonstrated ability to create a stable of entrepreneurial steeds through the Ain Center for Entrepreneurship at the Univer-sity of Rochester, comprise a distinguished and impactful legacy. He has done great honor to his family, notably his grandmother, Dora, through his accomplishments. And, those of us who have had the opportunity to intersect with him in our professional lives, will forever be grateful for his desire to make a difference and to strive, as per the University of Rochester's Latin motto "Meliora," for "Ever better."

Duncan Moore
Vice Provost for Entrepreneurship
Rudolf and Hilda Kingslake Professor in Optical Engineering
University of Rochester

Mark Zupan
President, Alfred University (2016–)
Dean, Simon Business School, University of Rochester (2004–2015)

PREFACE

Funny how certain moments in life create indelible memories. More than two decades hence, I can still see Kronos Incorporated founder and then-CEO Mark Ain come sauntering into the ballroom of a major hotel in Orlando, Florida resplendent, as he often was, in a brightly colored Hawaiian-styled shirt and Birkenstock sandals.

"How this?" he asked me with arms held wide to display what he clearly perceived to be sartorial splendor.

"What do you mean?" I asked back. "Is that what you're wearing for your speech?"

By that point I was already overloaded with anxiety. We'd shot interviews with Kronos customers the day before, then toiled overnight to create seven short videos that would play during this upcoming session, kicking off a three-day conference designed for an audience of about 1,500 Kronos customers. We had video introductions for each speaker and several other elements that made this a decently complicated two-hour endeavor. So, the butterflies in my stomach already threatened to lift me into the air like a real-life version of Tinkerbell.

This event would mark my sophomore effort as a freelance writer-director-producer, the basic role of the man-behind-the-curtain, for Kronos, the first having been an internal event held in Dallas a few months earlier. During that event, I'd worked with key executives, but this user conference marked my first outing with Mark, definitely the most free-spirited executive with whom I'd worked to that point in my career.

I'd already come to enjoy Mark. He was definitive in what he wanted. But he also took input. And you had to admire him for what he'd done to bring Kronos from start-up to a point where his company had been able to attract this many customers to Orlando that fall.

But Mark, at that time, had come to believe that he was a better speaker if he did not rehearse, a belief that I, who had been working on executive presentations for some time by then, did not share.

"Mark will not practice no matter how many times you ask him," his equally affable brother and Kronos Chief Operating Officer Aron had maintained enough times during our planning sessions that I finally and belligerently bet him $20 that I would ultimately coerce his elder brother to rehearse at least once.

At that moment in that ballroom however, my main focus was not on the words to be spoken . . . it was on the shirt.

"You can't wear that," I told him, perhaps more emphatically than prudent, given that several of the other speakers, Aron included, were within earshot.

"Why not?" he asked somewhat incredulously.

"Because, Mark," I said, "You're about to have 1,500 of your biggest and best customers walk in here and you should be wearing something more official like everyone else."

"But," he countered, "I didn't bring any other shirts down here with me."

"The Kronos store is out in the foyer," I replied, alluding to a practice whereby the company sold various logoed tops at these events. "You can grab a logo shirt out there."

"But," he countered, "I don't have any cash on me."

Once I explained to Mark that, as the CEO, it was unlikely that a Kronos employee would deny him a shirt on credit, he left the ballroom. But upon his return a couple of minutes later, he posed another problem.

"I can't wear this shirt," he declared while holding out an IZOD-styled top emblazoned with the Kronos logo. "The back is longer than the front. It will look silly on me."

The fact that a guy wearing a Hawaiian shirt was now worried about looking silly was lightly comical unto itself. But this was not the time to debate that point.

"I assumed you'd tuck it into your pants," was my reply.

"I can't do that," he said, "I don't have a belt."

I looked down. True enough, Mark's khaki trousers were without a belt. Needing a quick solution before the doors opened and the audience filed into the ballroom, I reached down, unbuckled my own belt, pulled it off my pants and held it out.

"You've got one now, brother," I said, inwardly fearing I'd overstepped my bounds but committed to finishing this play.

He looked at me. I looked at him. Mark took the belt, undid his pants (note that doors were not open for attendees yet!), changed shirts, buckled my belt, and again held out his arms.

"Anything else?" he asked.

"Yes," I told him. "I need you to go up on stage and read through your script one time before the doors open."

"Okay," he replied after yet another brief pause. "I'll do it."

I stood there thanking my lucky stars that Mark had allowed me to be so bold.

Then, as Mark made his way up to the ballroom's stage, Aron sidled up next to me.

"He's not going to improve his delivery by practicing just once," said the Ain brother that I knew had practiced his talk at least a dozen times.

"I know that," I said to Aron. "But now you owe me $20!"

The session went well and was positively received by the attendees. And I, in various capacities, have been working with Kronos during its climb from nearly $100 million in annual revenue to an industry-dominating global workforce management and cloud computing company that brings in billions a year, ever since.

In fact, I recently learned that I'm the longest-tenured vendor in the company's history. Sadly, that only tells me that I'm old.

Over the course of my involvement with the company, I've been struck by how successfully Kronos has navigated both the charted and uncharted business waters it faced . . . how the company expanded, and improved, and ultimately became the leader in an industry that Kronos both refined and redefined.

Along the way, I've been blown away by the overall quality and longevity of the people who make up the organization . . . good, hard-working, driven and friendly folks who serve as a key reason why Kronos is regularly cited as one of the best companies in the world for which to work.

The success is well documented. But through these years I've often sought to document how Mark got it all going forty-plus years ago . . . the motivation that drove him . . . and the somewhat nonscientific way (at least early on) in which he managed to create a core team that would stand the test of time and push the envelope further and further.

I'd spoken openly with Mark about this often off and on. But the timing never seemed to be quite right . . . until my phone rang over a year ago.

"Jim," he said after he'd asked about my wife and family, "I think I'm ready to tell the story."

But Mark being Mark, he didn't just want to sit down and tell things from his perspective. Rather, he wanted to share the experience with as many of those key players as possible, because he recognized that though he started as a force of one seeking to start a company, it was the collective power of many people who made the dream that became Kronos a reality.

Since that phone call, and based on the schedules of many good people, Mark has both told his story and acted as talk show host in discussing the perspectives of others. The resultant pages that follow are details of how things got up and running faster and faster as well as forays into the remembrances and contributions of those interviewed.

We won't be able to break down every twist and turn toward what is today a multi-billion-dollar organization. Rather, each chapter delivers a bit or a piece of what is ultimately a much larger puzzle.

Hopefully, for aspiring entrepreneurs out there, this represents a basic tenet that not only can you follow your dreams, but you can also do so and retain who you are at your core—even if you're the type of person who relishes wearing Hawaiian shirts!

Two things should be noted, by the way. The first is that Mark learned the value of practice and has become an increasingly better speaker over time (in fact, one of his speeches is included in this tome). The second is that, all these years later, though frayed to the point of breaking, and sadly not quite large enough for me anymore, I still have that belt. To me, it's no longer a viable piece of clothing—it's a trophy.

ACKNOWLEDGMENTS

The Entire Ain Family

All Kronites/UKGers Past, Present, and Future

All Kronos/UKG Customers

And All the Dreamers Who Dare to Call Themselves Entrepreneurs

SPECIAL ACKNOWLEGMENTS

Aron Ain

Alice Ain

Carolyn Ain

Larry Baxter

Bill Breen

Glenn Buldoc

Bob Cohen

Mary Jane Conary

Pat Decker

Cheryl Ferruccio

Donnamarie Floyd

Laurel Giarrusso

Michele Glorie

Jim Kizielewicz

Larry Krakauer

Stan Kulfan

Paul Lacy

Donald Levy

Duncan Moore

Kevin Oye

Jack Rich

INTRODUCTION

In a global world, a lot happens all at once.

In Ontario, a nurse manager at critical care health center quickly finds a replacement for a nurse who is sick and unable to come in to work her shift. Seems easy enough, but the nurse in question has an assignment that requires certain levels of accreditation. So, a replacement must be found within a staff of 2,600 who has that same level of training.

Meanwhile, in California's Sierra Nevada Mountain Range, where skiing is both fantastic fun and big business, payroll is prominent. One of the area's major resorts has ramped up its diverse seasonal workforce to over 2,500 employees, and they play roles in everything from recreation to hospitality, from real estate development to food and beverage services, as well as retail sales. That means simultaneously adhering to a multitude of pay rules to avoid things going downhill.

Further south, in Mexico, at a massive automobile manufacturing plant, more than 15,000 employees hold coveted jobs where workers earn roughly three times the national average wage by toiling in the international automaker's second largest manufacturing facility. But those workers have families, so some are putting in for time off at the same electronic time clock they use to check in and out of work.

Halfway around the world and down under in Australia, executives at a beverage company, creators of some of that continent's most loved and iconic nonalcoholic thirst quenchers, are collecting valuable data analytics in an effort to maximize the output of their workforce.

Meanwhile, in India, an insomniac recent college grad stays up late filling out a job application online in the hope of gaining employment with a retail consumer electronics giant. An employer to over 3,000 employees, this company is the kind of employer that empowers its staff with user-friendly tools like self-service kiosks, so hours can be checked and time off can be scheduled with a few taps on a keypad, providing just the type of modern environment that makes this college grad want to work for them.

Back in the United States again, it was likely that the customer service rep for a nationally known retail organization, who'd just returned from break to tend the chain's help line, gave little thought to these concurrent global goings on. Rather, with achieving a high level of customer satisfaction as her primary motivation, she was focused on assisting a gentleman who wanted to exchange a gift—a holiday sweater that didn't quite fit properly. Truthfully, at the other end of the connection, Mark Ain was interested in getting a better-fitting sweater. But he was, almost out of force of habit, likewise concerned with the rep's work environment from a Human Capital Management perspective.

"She was very nice and efficient," Mark will later say, "and arranged for me to get the same sweater in the size I wanted. But, at the end of the conversation I had to ask her, 'By the way, how does your company keep track of your time at work?'"

It's a question Mark loves to ask of nearly everyone he encounters, from the TSA agents at the airport to the checkout people at the local supermarket . . . even to the medical team that performed surgery on him a few years back.

"I knew that they all used Kronos," said Mark. And he would. Forty-plus years ago, he founded and subsequently led Kronos Inc., today recognized as the worldwide leader in the Human Capital Management space and one of the most successful software companies on the planet.

That Kronos now serves over 40,000 customer organizations worldwide, including all the organizations mentioned previously, is a testament to an environment Ain consistently strove to maintain, one where over 12,000 worldwide employees adhere to the same principles as established decades back, when those first few companies saw the benefits of utilizing an electronic time clock to record the time and attendance of their most valuable, and most costly, resource: their employees.

"The poor service rep didn't know what to answer," Mark recounted with a twinkle in his eyes. "She told me she had never bothered to look at the system itself. But she assured me it was accurate and promised to check on who made it when she left at the end of her shift. 'I'm going to look this time,' she promised."

Mark didn't tell the rep that he was the founder of a company that sustained one of the longest records of growth and profitability in software industry history and earned its reputation as the market leader in workforce management. He didn't boast that this record of growth and profitability was bested by few organizations. Because that wasn't nearly as important to him as the fact that he was treated well by an employee who was focused on doing her job as best as possible. Oh, and that he got the right-sized holiday sweater.

Other business moguls, and Mark Ain is indeed a mogul by most measures, might have made a bigger deal out of this and so many other similar interactions. How could they not know Kronos? Why don't they recognize his name? But Ain is aware that even today, despite the fact that his organization empowers those 40,000-plus organizations to manage the full life cycle of over tens of millions of employees on a daily basis through a series of integrated solutions that run the day-to-day management of employees . . . even though Kronos essentially frees up management within those organizations to focus more fully on achieving their own company-based goals—be that through properly staffing emergency room and shop floors, accurately tracking and paying thousands of employees, or providing a seamless way for qualified candidates to apply for employment—Kronos, and the name Mark Ain, remain little known in most circles.

And that's okay. Because you need to understand, the low profile is, like most of the things Mark does, very much by design.

His has most often been a quiet and humane management style, one best reflected by a quote attributed to someone Mark has studied extensively: the ancient Chinese philosopher and writer Lao-Tzu: "A leader is best when people barely know he exists," the philosopher espoused, "not so good when people obey and acclaim him. Worse when they despise him. But of a good leader who talks little, when his work is done and his aim fulfilled, they will say, 'We did it ourselves.'"

"I wasn't always as easy-going, though," Mark confessed with a chuckle. "In fact, there were times early on when I wouldn't have wanted to work for me."

But in general, folks who joined his company stayed on far longer than most would consider the norm, worked hard, and prospered along with the organization as it grew from a tiny start-up in Massachusetts to a company serving customers in over 100 countries around the world—one that as of this writing generates annual revenues north of $3 billion dollars a year.

And if he was hard on himself and others in an effort to get the clock ticking on four-plus decades of success, the end-game shows that his method not only assured longevity for Kronos, it likewise cultivated a culture whereby Mark is able to proudly point to the multitude of times Kronos has been labeled a "Best Place to Work" and received scores of awards for excellence in customer service.

Long since retired from his position as the company's CEO, a role now held by his youngest brother Aron, Ain remained on the Kronos Board of Directors until 2020 and also currently serves on various other boards. And though he most often prefers to be anonymous and let others garner the accolades, his is a story of drive and ambition, of overcoming obstacles and building teams, of seizing the moments as they came, and of making the right strategic choices at seemingly the precisely correct times.

In other words, looking fresh in a perfectly fitting holiday sweater, Mark Ain has a story that needs be told. . . .

Kronos by the Numbers

How the growth of Kronos Incorporated from an idea to the basements and garages of its early core team to a soot-filled ironworks foundry to a multi-billion-dollar global leader in an industry that it refined, redefined, and ultimately led is documented on the following pages.

This story lies in the building of the company—based on an idea but founded on its people—that would stand the test of time. Founder Mark Ain took a nontraditional approach to team building that prioritized characteristics over resumes, talent over gender, and potential over past accomplishments.

Within these pages, we'll get a sense for the forces that shaped young Mark Ain as an entrepreneur . . . how his adventurous spirit and fearless nature in his formulative years readied him to be a risk-taker and entrepreneur.

We'll "listen in" as Mark looks back with many of those key contributors who shared Mark's technology-based belief that microprocessors would eventually change the world . . . and that the relatively mundane function of timekeeping was ripe for such an electronic boost.

And we'll get a sense for how the success of his company led Mark, the epitome of a successful business creator, to leverage his own success and take steps to foster a new generation of like-minded visionaries.

It should also be said at the outset that as the process of documenting those formulative Kronos years took place, the company went through a major transition when, in the spring of 2020, and amid a worldwide pandemic, Kronos merged with Ultimate Software to form a workforce management organization with over 12,000 employees worldwide. The combined companies have now formed a single organization rebranded as the Ultimate Kronos Group or UKG.

So, albeit within this new permutation, the name Kronos (the founder's second choice for a name) lives on within an organization that's gone from losing money to a near-record-setting performance for revenue and profitability growth—from no sales to a trickle to a worldwide leader with tens of thousands of customers and annual revenues measured in the billions (with both a "B" and an "S"!)

What the following story of Mark Ain and Kronos, Inc., proves is that guts and perspiration mixed with the right blend of the right people can equal success at a level beyond your dreams. It is our hope that this story will shine a light for future generations of entrepreneurs.

Chapter One

GROWING UP AIN

The journey of 1,000 miles begins with one step.

—Lao Tzu

The year 1943 was dominated by an ongoing World War II. The United States ran its first test on what was called the Philadelphia Experiment. And a young PT boat commander named John F. Kennedy bravely held his crew together in the Solomon Islands after their boat had been torpedoed by a Japanese submarine.

The ballpoint pen was patented. Construction on the Pentagon was completed. And Franklin Delano Roosevelt became the first U.S. president to travel by airplane.

In New York, Rogers and Hammerstein's *Oklahoma* debuted on Broadway. The New York Yankees became the first team to win the World Series ten times and Yankees star Joe DiMaggio promptly enlisted in the U.S. Army.

As for beginnings, and against the musical backdrop of upstart crooner Frank Sinatra singing "All of Nothing at All" and *Wizard of Oz* alum Judy Garland warbling "Zing Went the Strings of My Heart" came the births of such future luminaries as Robert DeNiro, Jim Morrison, Billie Jean King, Geraldo Rivera, Joe Namath, and a couple of British blokes named Richards and Jagger.

Future household names arrive on the planet nearly every day, but if history proves anything with consistency, it's that most world changers do not necessarily become world renowned. And such was the case when, in the year 1943 in New York City, Pearl and Jacob "Jack" Ain welcomed the first of what would be their five children, Mark Stuart Ain, into the world.

To imply that young Mark had the potential to be successful would almost seem an understatement. Jack had earned an engineering degree from New York University in 1937. But it was Pearl who truly had achieved eye-popping academic credentials.

"My mother graduated high school when she was only fourteen, then college at just seventeen, and she finished Columbia Law School at twenty," Mark reported with pride. "But she was too young to take the test! She had to wait until she turned twenty-one to take the bar exam!

"And so, she had a fabulous academic career. And then, she practiced law for a few years, I guess until I was born."

But there was no immediate mercurial rise for young Mark.

"I was a pretty sickly child," Mark recalled. "I developed a severe case of asthma when I was three or four years old and, from that point forward, I was in and out of the hospital quite a bit. I was so skinny and I was not in good health. And if you see pictures of me at a young age, I look like I was starving."

"When he was young," the late Jack Ain would recall during a taped interview, "he had very bad asthma. So, I set him up in a room with an air conditioner and a dehumidifier. So, he spent a lot of time in his room."

"He was a very independent child," Pearl Ain chimed in. "He liked to stay by himself and read and do things on his own. His personality was such that he always wanted to develop his own agenda and follow it."

Not that young Mark was overly shy or didn't like to interact with others.

"He made friends easily," said Jack. "He was very outgoing. But Mark was also always focused on what he wanted to do. He could never get sidetracked."

"I think he always had a dream that he wanted his own business," added Pearl. "Fortunately, over time, his dream was fulfilled."

Soon, the burgeoning Ain Family, now five strong with siblings Brent and Ross joining Mark, moved to Jack's childhood hometown of Sea Cliff, New York, so the elder Ain could take over his father's plumbing and heating supply business. Pearl, as was the mindset of the time, had stepped away from the law and become a full-time mother as Mark, Brent, and Ross were later joined by Aron and Alice.

Luckily, shortly after the move to Sea Cliff, Mark enjoyed a breakthrough of sorts.

"At some point, I don't remember when," said Mark, "I started going to an allergist and getting weekly and then biweekly and monthly shots. And that changed my life because these treatments allowed me to become part of the regular world. Getting the asthma under control . . . that was a huge step."

The move, and his under-control-but-never-gone medical adversity, would shape Mark's life in a couple of ways.

Located 25 miles east of Manhattan and sitting atop a 120-foot bluff on the North Shore of Long Island, Sea Cliff was a tiny hamlet nestled along the shoreline of Long Island's renowned "Gold Coast." Though small in size at only one square mile, Sea Cliff nevertheless boasted its own beach and sixteen neighborhood parks, and also bordered the Harry Tappen Beach. With its breathtaking views of Hempstead Harbor, Long Island Sound, and the New York and Connecticut shorelines, Sea Cliff was a natural playground for the young Ain brothers.

Life was good, but challenges remained.

"I ended up going to a very small-town elementary school," Mark recounted. "But I was not very well. I always got sick more than other people because of my bronchial tubes, which had been weakened by the asthma. And I was pretty small. So naturally, there was one guy, and he later went to Harvard, who used to beat me up most days walking home from school. That wasn't very pleasant. But it forced me to develop a sort of toughness.

"And," Mark added, "because I was still so sick, I continued to spend a lot of time reading books and doing things by myself."

The book reading in particular would prove fortuitous. With no cable TV, video games, or the internet as distractions, Ain pored over books and excelled in his early studies so that, when Sea Cliff merged its school system with two other neighboring communities to form what is even today regularly rated as one of the best public school systems in the country, young Mark Ain moved quickly to the top of the class.

"I went, starting in junior high school, to the newly formed North Shore School District, which was a combination of three districts," Mark recalled. "And, by the time I was in sixth or seventh grade, my asthma was sufficiently under control so I could play sports. I played some intramural soccer, that sort of thing.

"In high school, I think there were 200-plus people in my class. So, it was a relatively small high school. But I was the star in math and science. In fact, when I was a sophomore, I proved that the math teacher, who had been teaching at North Shore for forty years and had all the answers written down in his book, I proved to him on a few occasions that the answers, that he'd been using over those forty years, had been wrong," Ain laughed. "He didn't like me very much."

Along the way, Mark found himself a bona fide hobby. He took up photography. This passion led him to be a contributor to the school newspaper and the yearbook, and inspired him to create his own dark room in the family basement, where he developed film and created custom prints. It was his first foray into creating a working environment, albeit one best suited for a single employee.

Happily, Mark also discovered a high school sport that would test his discipline and training without regard for his smaller stature.

"He was on the wrestling team," Pearl proudly remembered. "He frequently had to use an inhaler to control his breathing, but he never gave up. He wrestled in spite of his disability because he wanted to do something in athletics but he wasn't tall or big enough to do many of the other sports, and wrestling was something you did in your own weight class. So that is what he chose."

"I was on the wrestling team all through high school," said Mark. "I wrestled in the 134-pound weight class. And you'd think there wouldn't be too many boys that small who wanted to wrestle. But there was another guy and I who were in the same weight class. Once a month we'd wrestle off to see who was tops in the weight class and the victor would then wrestle for the varsity. So, we had our own internal competitions, plus battles against wrestlers from other schools."

While in high school, new siblings came along to join Mark and his two brothers. First came Aron, born in December of 1957. And then Pearl finally got her girl, with Alice following thirteen months later, in January of 1959.

"Five wasn't considered a lot of kids back then," Mark remembered. "I'd say, in general, we had a pretty good family."

In terms of influencers, family played a role in the young Ain's development.

"My father was not a great businessman," said Mark. "He was a kind person who had great stamina. But rather than putting the business first, he took on many outside responsibilities. He would be out doing some sort of charity work almost every night of the week. So, we didn't see that much of him because he was always, not minding the store, but doing all these other things. He never really could make the business go, because he was too distracted. At the same time, he was not a delegator. Nobody else could do the billing correctly except him. So sometimes he didn't get around to billing people for ten years. Now, if you're a plumber, are you going to pay a bill that's ten years old?"

Jack was not without a way of teaching his young sons the value of hard work, though.

"My father had a big maple tree cut down in our yard," Mark remembered. "But rather than spend the money to have the stump and roots taken out professionally, he decided we boys would remove it."

The combination of the size of the tree and the relative smallness and youth of the stump removal workforce meant the Ain brothers had a project on their hands that would ultimately take months.

"A lot of the time, allegedly to help us out, my father would pour water from the garden hose into the hole we were digging," Mark laughed, "so we'd be miserable, digging in the mud. But I think we learned the lesson he was trying to convey. We each committed to doing well in school so that we wouldn't have to be laborers later in life!"

Mark also credits his grandparents for inspiring him and his siblings.

His paternal grandmother had passed when he was very young, but his grandfather, his father's father, a retired entrepreneur, used to come over to the house almost every day. And the elder Ain would take him fishing, and help him buy wood to build things like a desk.

"I enjoyed my time with my grandfather," Mark said in retrospect. "I think he was trying to fill in for things my father didn't do because he was always on the go. But I'd say it was my Grandmother Ponemon, my mother's mother, who really had an impact on me in terms of how I eventually decided that I had to become an entrepreneur . . . my own boss."

As the story goes, when Dora Ponemon was twelve or thirteen years old, her husband-to-be Harry Ponemon left Grudno, a city at various times annexed by Russia, Poland, and Lithuania, and today part of Belarus, and emigrated to the United States. Several years later, when Dora turned seventeen or eighteen, Harry sent her money for a ticket to come to America to be his wife.

"When he sent the money, he included a note that said, 'If you don't want to come, you can keep the money,'" Mark recalled.

Dora first took a train to Hamburg, then a boat to Liverpool, and finally a ship to New York City. But rather than rush to Harry, she went to stay with some relatives in Brooklyn after she passed through Ellis Island.

"My grandfather came to see her and asked, 'What are you doing? Aren't we getting married?' to which my grandmother asked, 'What do you do?'

"Harry said, 'Oh, I have a very good job. I earn good money. I'll be able to make a good living for you and our children,'" said Mark.

But that reply was not good enough for young Dora. She pressed Harry further, asking him exactly what he was doing to earn a living. A befuddled Harry tried to explain how his current position held great potential, but she cut him off with an analogy that would become folklore within the Ain family.

"She put it right to him," Mark recollected. "She said, 'I want to marry a horse, not a mule. Mules work for other people. Horses run their own companies. I don't want to marry somebody who works for somebody else. I want to marry someone who does his own thing . . . someone who has their own business.'

"And so," Mark chuckled, "my grandfather acquiesced and, soon after they were married, my grandparents started a haberdashery business together. My grandmother would mind the shop with the three kids, my mother included, and my grandfather would go out measuring and installing drapes and sofa covers."

Through hard work and some luck, the Ponemons became so successful that Dora took some of the profits and started buying apartment buildings in New York City.

"In the end, I believe they owned four apartment buildings," said Mark. "So, they did, indeed, become horses."

Unfortunately, like his father's mother, Harry passed away when Mark was very young. But Dora, still a thoroughbred later in life, was able to manage those apartment buildings in the Bronx, while spending about half of her time living with the Ains on Long Island.

During that stretch, Mark, as the oldest sibling, was tasked with driving Dora between the Bronx and his parents' house. And it was on those many drives that she would stress to him the importance of being a success.

"I really credit her for a lot of my entrepreneurial drive," said Mark. "She was an enormous influence for me, all through growing up. She had been a huge success in America, basically through sheer force of will."

And despite his humble, asthma-dominated youth, Mark was an early success. His photography skills were lauded. He was respected for how he approached his challenges on the wrestling mat. He was in the National Honor Society. He was the captain of the Math Team. And academically, he was at the top of the heap in a highly competitive school. Which introduced the question . . . what would he do upon graduation?

"I was told I was the smartest kid to go to my school in math and science," he recalled. "And it was stressed to me that nobody from my school, in forever, had gone to MIT (the Massachusetts Institute of Technology). And so, they urged me to apply to MIT."

"His grades were very good," father Jack Ain would say. "And somebody said he should go to MIT. So, we went up to MIT and we were very much impressed with it. So, we agreed with the school and encouraged him to apply."

High aspirations—the Massachusetts Institute of Technology was then, as it is now, considered the world's preeminent engineering university. Justifiably, the reasoning holds that it is, likewise, one of the most difficult institutions of higher learning to which to gain acceptance. Undeterred, the young man who had overcome medical hurdles, the occasional bully, and the rigors of studying his way to the top of his class completed his application. And not long thereafter, the mailman delivered the answer, not in a thin envelope, but in a thick registration packet.

"I decided," said Mark, "to go to MIT."

Chapter Two

AN EDUCATIONAL ODYSSEY

Getting an education from MIT is like taking a drink from a fire hose.

–Jerome Weisner, Former MIT President

Young Mark Ain arrived in Cambridge, Massachusetts, and the Massachusetts Institute of Technology in the tumultuous fall of 1960. Vietnam War protests would be rampant in a few years, but right now citizens were focused on the recent discharge of one Army sergeant, a Tennessee native named Elvis A. Presley. The Space Race was in full swing, as dreaded Cold War foe Russia sent Sputnik 4 into orbit. And there was a well-documented voyage to the bottom of the sea as explorers Jacques Picard and Don Walsh descended nearly 36,000 feet below the surface of the Pacific Ocean to the bottom of the Mariana Trench.

Massachusetts senator and war hero John F. Kennedy was squaring off with Vice President Richard M. Nixon as the 1960 presidential election turned down the back stretch. The Reverend Martin Luther King Jr. fired people up with his rhetoric as the Civil Rights Movement pressed for equality for all. And a brash eighteen-year-old Cassius Clay captured the imagination of the country and the Olympic gold medal in heavyweight boxing.

The aggregate revenues generated by companies founded by MIT alumni would today rank as the eleventh largest economy in the world. Companies like McDonnell Douglas, Texas Instruments, 3Com, Raytheon, Campbell Soup, Intel, and even Koch Industries each shared a common bond. They, among tens of thousands of other organizations, were all founded by MIT alums, resulting in the employment of more than three million people and generating collective annual revenue north of a staggering $1.9 trillion. For a single university, that is nothing short of mind-boggling. But that didn't matter too much to Mark Ain after he arrived in Cambridge and checked into his freshman dorm room that September. He had far bigger concerns.

"I soon discovered that I was not the smartest person ever born in math and science," he recalled. "And, truth be told, I struggled to keep up right from the start."

"Mark overcame great difficulties in his first year at MIT," mother Pearl recalled. "He hadn't had calculus or any of the advanced placement courses because our high school didn't offer them. So, when he got there he was at the bottom of the heap, and he had a great deal to overcome."

Having been the proverbial big fish in what turned out to be a relatively small North Shore High School pond, only to struggle to swim in these academic waters was disconcerting, to say the least. But even though North Shore hadn't quite prepared him for the full rigors of the world's preeminent engineering university, there was solace and support connected to that North Shore experience.

"I looked up in those early days, and there were a couple of familiar faces," Mark remembered.

Those faces belonged to a couple of fellow Long Islanders whom Mark had encountered in high school during his interscholastic math team competitions.

"Our math team was in a league with, I seem to recall, six other schools. Truthfully, maybe those competitions should have told me something because I think we always finished last," Mark chuckled. "But two of the captains of the better teams were also at MIT . . . really brilliant guys. And one of them was Donald Levy, who was in my dormitory. He and I became friends."

Not that meeting Donald transformed Mark into a top student at MIT. Quite to the contrary, Donald was a tough act to follow, finishing all but two courses and his thesis in his first three years. But the friendship, which would later manifest itself professionally, gave Mark someone to relate to as he dug deep and did what he had to do to keep pace with the rigors before him . . . at least to a degree . . .

"Donald was a great friend. And we are still friends to this day. I'm not sure he ever got anything but an A, whereas I struggled to get by," Mark confessed. "Plus, away from home for that first time, I also discovered girls, which meant that, quite quickly, there were a lot of things that I was doing that were far more important to me than academics."

It didn't help that his major was electrical engineering, at the time and still today one of the most difficult majors, and at one of the most difficult schools on the planet to boot.

"I persevered," he continued, "but I questioned my path, especially after fourth term physics. In that course, 50 percent of the class ended up getting an F because the coursework was impossible. And me? I was one of them."

That this particular class was so difficult still brings a smile to Mark's face, if only because it was taught by yet another future technopreneur, as MIT likes to call many of its alumni, one Dr. Amar Bose, who would go on to found the Bose Corporation and forever change the nature of how the world records and listens to audio, in 1964.

Accepting of his unacceptable grade, but unwilling to fail at his quest for an MIT degree, Mark slipped out of Cambridge and into a summer school class on Long Island, where he passed that required level of the Physics curriculum.

"Eventually, when we were juniors, five of us, including my friend Donald Levy, moved out of the dormitory and took an apartment in Harvard Square," said Mark.

The young men enjoyed the sense of autonomy that came with living off campus, as well as the overall vibe of Harvard Square. To add to that freedom, Mark and another roommate pooled their money and purchased what many would call a "land yacht," a massive pink and white 1956 Buick convertible that was roomy enough to carry them all, and big and bad enough to "create" valuable parking space on the space-deprived streets of Cambridge. The purchase price was a whopping $150, but the intrinsic value was priceless.

"No space was too tight," reported Mark, "because with this car we could just push other cars out of the way and make space."

With good friends sharing common educational and life experiences, the workload remained hard, but the daily trek on the proverbial path up and back on Massachusetts Avenue, whether in the Buick or on his trusty bicycle, became incrementally easier.

"It didn't make the classes less challenging. In fact, I struggled with other classes, and at various times I really didn't think I would graduate. But it taught me to balance hard work with having fun," he said. "So, I learned some valuable life lessons to go along with all of those difficult classes."

Not all who entered the Massachusetts Institute of Technology in the fall of 1960 persevered, or, more accurately, survived. A good portion less than those who started that year succeeded. But, in the spring of 1964, Mark Ain heard his name called as he strode across the stage in front of family and friends and accepted a hard-earned degree in electrical engineering.

But that didn't mean he was then ready to step out and conquer the world of business by redefining an industry through the harnessing of technology. Quite to the contrary, and likely against what was expected by many in the audience that culminating May day, Mark knew that what he needed to refuel his personal fire was an extended break.

In Australia, the aboriginal people call it a walkabout. In Native American culture, it's called a vision quest. Mark didn't put any particular label on his decision to go on what would be termed a sabbatical from following some traditional path. He simply sought a change of pace . . . and a change of venue.

"I decided I wanted to leave everything I knew behind," he said. "So, I worked two jobs in Cambridge for a month and saved every penny. I combined that with some money that my grandmother had given me over the years. And I took a boat to Le Havre in France."

Sounds like an extravagant way for a recent college grad to journey to Europe. But when the Beatles and Rolling Stones invaded America that year, they did so by the newest form of popular travel, via one of the increasing numbers of trans-Atlantic airplane flights that forebode the end of the golden age of transatlantic oceanic travel via ship.

So, in late June, Mark boarded the Cunard Line's half-filled *Mauretania*, the little sister of the mega ships of their day, the *Queen Mary* and the *Queen Elizabeth*, in New York for a more than reasonable price armed with only a suitcase and a piece of paper with the names, addresses, and phone numbers of some European acquaintances he'd met the previous summer.

"I'd taken a job as a chaperone on a bus trip around the United States in the summer of 1963," he said. "And a lot of the students were from various European countries. We'd become friendly. They probably thought they'd never see me again. But I got on that ship knowing I had a lot of people to visit in Europe."

Mark arrived in France a week or so later and took a train to Paris, where a series of Bastille Day celebrations began what would become a seven-month, unplanned, unscheduled, and completely underfunded trek through and between countries that were completely foreign to him.

Over a decade later, Mark would become known for innovation through technology. But for this trip, innovation became his modus operandi where travel was concerned.

"*Europe on $5 a Day* was my Bible," he recalled.

That meant traveling light and cheap. So once the celebrations ended in the City of Lights, the recent MIT grad broke out his thumb and hitchhiked, first to Frankfort, Germany, then to Hamburg. From

Hamburg, the thumb then took him to Copenhagen, Denmark, where he met up with a friend from high school.

"When it was time to move on from Denmark, I decided we should upgrade our transportation," he said, "so I used some of my travel money and bought a used Lambretta motor scooter and, in the pouring rain, we rode to Stockholm, Sweden."

Stockholm's port on the Baltic Sea was spectacular. But once he'd dried his clothes, it was another 250 miles across Sweden to Trollhattan, best known as the location of Saab Automobiles. Mark was not so much interested in these safety-oriented vehicles as he was in the daughter of a Saab executive, a young woman named Aina who was yet another of the students he'd met during that fateful bus trip the previous summer.

"I fell in love," he remembered of the whirlwind romance that budded during a two-week spin from Trollhattan to the family's second home, another 50 miles away on a fjord in Gothenburg.

But the lure of his personal tour sent him back on the road.

"After two weeks I got back on my motor scooter and, after driving through the mountains of Northern Switzerland, I ended up in Northern Italy," he said. "I left the scooter with yet another woman from that same trip because I'd decided to hitchhike to Istanbul . . . except, in Zagreb, Yugoslavia, I ran out of cars!"

Mark learned that the Orient Express would be coming through Zagreb in the middle of the night. Even in 1964, boarding a train at that time was not the best thing to do when one was traveling alone.

"It was after 1 a.m. and this group of guys at the train station decided they would beat me up, rob me, and take my passport," Mark recalled. "It was scary."

Luckily for Mark, he looked toward the end of the platform and mistakenly thought he recognized a drinking buddy from his time in Amsterdam a few weeks earlier.

"I called out to the guy by name and he came over," recounted Mark. "But as he got closer, I realized it wasn't the man I'd met in Amsterdam."

Mark can list many times in his life when luck has played a role in his success. But this instance stands out. Some 800 miles from the capital of the Netherlands, he had mistakenly called out to someone who only looked like the man he remembered. But the young man on the platform, a student on break from studying in Russia, turned out to be a compatriot and friend of the man Mark thought he was calling to rescue him.

"He was a large man," said Mark. "And he knew my friend! So, he told the group that I was with him. They took a good look at him and decided they would leave me alone."

With the fear of physical harm cast aside, the odyssey continued. Mark made his way to Istanbul where he reveled in the culture until he woke up one morning in a youth hostel in the midst of an asthma attack.

"I wasn't ready to go home just yet," he said. "But I do recall that incident made me take account of what my longer-range plans were. I wondered to myself 'What am I going to do when I get back home?' And that day I decided that I would pursue an MBA."

The revelation did not send Mark scampering for the states. Rather, he concluded his time in Istanbul, boarded yet another ship and crossed the Aegean Sea so he could climb the Parthenon in Athens. He then jumped back on board and the ship took him to Florence, Italy, where his burgeoning hitchhiking skills took him to Rome, where he met up with a woman who'd been a law school class-mate of his mother's. Then it was back to Northern Italy and his trusty scooter.

"I took the scooter back to Paris," he said, implying that this was no big deal (and likely by this point it wasn't), "and I stayed there for four months except for a two-week break during Christmas, when I went back up to Sweden to see Aina and spend the holidays with her family, eating the most delicious smoked fish."

All good things must come to an end, however, and this unplanned and itinerary-less trip was no exception. And so, after a journey of thousands of miles, visiting many countries and seeing as many peo-ple as possible, given his recent college grad financial resources, Mark bought himself passage on another transatlantic voyage and returned home to New York in February of 1965, rejuvenated and ready to forge ahead with his life.

"I got a job in New York City and started applying to graduate schools," he said of his post adventure life.

However, with one sibling in college and another slated to enroll the following fall, Mark's father made it clear there was no money to finance his MBA ambitions.

So, with money low, and choices limited, Mark's decision became an easy one.

"The University of Rochester, which was a fantastic school anyway, this would be the first year they were offering an MBA program," said

Mark of the school where he has now had a decades-long relationship. "So, they were looking to attract students and offered me a scholarship and a monthly stipend. And one of my roommates from MIT was already there working on his PhD in economics, so I had somebody to live with. So, it was an easy decision."

That easy decision led to a couple of fortuitous ones. Shortly after arriving on campus, Rochester's fledgling MBA program aligned itself with a French-based, student-run organization called the International Association for Students Interested in Business and Economics (AIESEC), an international association for students interested in business and economics. The group's stated mission was to develop business leaders throughout the world by facilitating international exchanges between businesses and students. One of the group's first UR recruits was recent international traveler Mark Ain.

"Our charge was to call on senior management within companies in upper New York State to convince them to bring trainees in from around the world," reported Mark. "And for every trainee position we opened up for the organization, a position would be offered to a Rochester MBA student abroad."

Mark met with heavy hitters like the respective CEOs of Xerox and Kodak, and was successful not only in securing trainee positions at those organizations, but also in recruiting these powerful individuals to serve on the Rochester chapter's Board of Directors.

"I learned a few things in the process that would serve me well later in my career," he said. "First of all, I learned that even though these were very high up, brilliant people, they were just human beings and, for the most part, good people. But I also learned that going out there myself and calling on these people cold . . . if I could do that, I could make a sales call anywhere."

Ever the competitor, Mark was especially gratified that the effort put forth by him and the rest of the Rochester grad students resulted in their university being recognized as the second most prolific collegiate effort in AIESEC in the country during his time there. And he saw that second-place finish as victorious for one important reason.

"We finished second to only Yale University," he said. "But the students at Yale didn't have to go out cold calling like we did. All they had to do was ask their fathers to create positions for them. We had to work for every position we secured."

His second year afforded him more training as he took interested students under his wing and on the road to meet with corporate executives.

"So, my first year I was mainly selling whereas my second year I was teaching people who were in their first year how to sell to companies," he recalled. "That was also a great learning experience."

In the summer before that second year, ever seeking opportunity, Mark took one of those overseas trainee positions for himself, venturing to Birmingham, England, to research expansion possibilities for a printing organization called Kalamazoo Limited.

"They manufactured various kinds of paper," he recounted. "My role as a student that summer was to help them figure out what other countries to expand into. I went to London a lot to do research . . . remember there was no internet so research meant digging in and doing due diligence. But it was also my first real experience consulting, which was helpful later on in my career."

Of course, Mark remembered more than just the job.

"I lived in a house with a group of students from all over the world, in the worst section of Birmingham," he laughed, "and Birmingham is pretty bad to begin with. The place was so bad that I took my showers across the street at the athletic club!"

But it was dealing with experiences like his Birmingham summer, under sometimes difficult conditions, as well as his first year with AIESEC that sparked a revelation at the end of spring that year. Mark was able to identify a key fundamental strength. True, he was intelligent. And true, he was able to adapt to varying forces around him. But at his core, Mark came to realize that he was a team builder.

"I decided to major in organizational behavior," he recounted. "So, my focus was on how organizations work and how to best manage organizations."

Not only had he discovered a pathway by which he could become successful, he also found a mentor.

"Organizational behavior was taught by a professor named Ed Henry, who had just retired from a position as the vice president of Human Resources at Esso," said Mark. "He came to Rochester in my second year and he taught a course both terms.

"But," Mark continued, "we as students really learned about business by going to his house every week for a potluck dinner."

As the former head of Human Resources at Esso, today known as Exxon Mobil, Henry had developed quite a list of influential and knowledgeable friends, top executives from companies like General Electric and General Motors, many of whom joined the students for those weekly meals.

"We went over to Professor Henry's house every week and met with some of the top business leaders in the country. In that casual environment, you heard all kinds of things that you couldn't learn from books," Mark recalled.

The professor, and his informal dinners, had further positive influence as Mark chose a topic and worked on his Master's thesis.

"My topic was called Leadership Group Discussions," he said. "And it was really an offshoot of those dinners. So, I'd get a bunch of students in a room and I would tape-record the discussions they had on different business topics. Then I'd bring in another group of students to listen to the recordings and have them evaluate the various students who been recorded . . . who sounded like a good leader . . . who was a visionary . . . that sort of thing."

Courses completed and thesis graded ("I'm pretty sure I got an 'A'"), Mark's family had the dual honor of running back and forth across the campus at Rochester as Mark received his MBA and brother Brent was handed his undergraduate diploma on the same day!

"That was quite a day," Mark laughed.

Mission accomplished; Mark was ready to travel again. But this time his plan was much more solid than his previous one-man Amazing Race. Through the same International Association for Students Interested in Business and Economics, he'd lined up consecutive assignments. First, he'd scored a five-month stint in Tokyo, Japan, beginning in the summer, working for the Japan Tobacco and Salt Public Corporation, a complete state monopoly under the direction of the Japanese Ministry of Finance. From there he'd travel to Sydney, Australia, where he was signed on to work for a technology company that would have him, in Mark's words, "working near the beaches."

Armed with a world-class education and set for a worldly work adventure, much seemed locked in place for Mark Ain. But, before he could leave New York yet again, a letter arrived in the mailbox that would forever alter his meticulously preplanned trajectory.

Chapter Three

WORKING UP TO AN IDEA

Let reality be reality. Let things flow naturally forward in whatever way they like.

— Lao Tzu

The year 1967 would later become known as the Summer of Love, but love wasn't all the world needed in that fateful year, not with the Vietnam War in full swing.

The specter of this long-running and unpopular conflict, that sent young men (and some women) halfway around the world to wage war against people most Americans knew little about, hung over everything like a proverbial American flag obscuring any view of the coffin of a fallen American soldier.

Protests became the order of the day. Returning veterans, already dealing with what would one day be called post-traumatic stress disorder (PTSD), were spat upon rather than revered. And, in an act of civil disobedience, the former Olympic gold medal hero and current boxing heavyweight champion of the world, the former Cassius Clay who was now known as Muhammad Ali, was stripped of his title and banned from boxing for refusing induction into the U.S. Army.

Mark Ain, fresh off his own successful endeavor to earn an MBA with a concentration in Organizational Behavior, was set to travel to Japan for a work assignment he'd set up just prior to receiving his advanced degree.

Once his Japanese assignment was completed, he was already set up to journey to Australia, again for what would be valuable international work experience (though, truth be told, the fact that he'd be living and working on or near the beaches held measurable allure for Mark as well).

And, further toward the horizon, Mark had lined up full-time employment on the West Coast, in San Francisco, where he planned to settle.

But just as it was time to pack his bags and embark on this personally designed global indoctrination into the world of business, the Vietnam War arrived in his mailbox. Mark Ain had been summoned for a draft physical.

"I'd lined up a job in Tokyo for five months. I'd already set up a second assignment working for a technology company in Sydney that would begin after that. But just as I was getting ready to leave for Japan, I got the letter," he detailed. "That was the Vietnam Era . . . if you weren't in school, you got drafted."

Now no longer a full-time student, Mark was put on notice that he would be receiving further instructions. More immediately impactful, the designation came with travel restrictions that negated his ability to fulfill either the Japanese or Australian assignment. Said restrictions had been instituted as a deterrent for potential draft dodgers by the U.S. government.

"You couldn't leave the country for as long as I wanted to leave it, so I was essentially stuck," he said.

Stuck in more ways than one. Believing that these out-of-the-country jobs would give him valuable experience, Mark had turned down an excellent job offer with Esso Chemical that he'd secured through former Esso executive Ed Henry, his Organizational Behavior professor at Rochester.

"I called Professor Henry and told him that I couldn't leave the country. I asked if the position was still open. But it was too late. That job was gone."

Luckily in a multitude of ways that would become apparent later, Mark was told that, though the position he'd been offered had been filled, there was an opening for an MBA at Esso International, which at the time was the operating arm of the Esso Corporation that oversaw the increasing flow of ships, particularly oil tankers, in the Persian Gulf.

"So, I went to work for Esso International in Rockefeller Center in New York City," Mark detailed. "And my job there was routing ships and tracking where they were going after they loaded up with their cargos."

As they have seemingly been for all time, things were particularly dicey in terms of cargo ship travel that Summer of Love. Nearly simultaneous to Mark taking his position at Esso International, the 1967 Arab-Israeli War had broken out. Though it officially lasted just six days (and also came to be known as the Six-Day War), this conflict between Israeli forces and the combined forces of Egypt, Jordan, and Syria over control of the Sanai Peninsula, the West Bank, and

the Gaza Strip almost literally wiped out the Egyptian Air Force and
brought calamity to the region well beyond the documented duration of the clash.

Over 5,500 miles away, Mark had little time to focus on either
the fact that he had been prohibited from going to Japan, or even
his draft status. He needed to learn the ropes quickly for fear that
a ship would be attacked or hijacked. And the order of the day was
misdirection.

"As a ship left port, we would tell them they were going on one
route," he explained. "Then, as soon as that ship would get to a certain
point, we would reroute it so that it couldn't be intercepted."

The result was safer passages for the ships—and a temporary safe
harbor for Mark as, upon review, the U.S. government saw a need to
not draft him after all, as the safe movement of those cargo ships was
critical to U.S. interests.

"I became essential to U.S. security. It was not planned that way. It
just happened," he chuckled. "They couldn't have me come in for a
physical because it was decided that this job I had was too important to
allow me to be drafted."

The reprieve was relatively short-lived however, as highlighted by
the conflict's Six-Day War moniker. That's when an old hindrance
became an unexpected ally.

"By the time that short war ended and things started settling down
in the shipping lanes, I got another notice that said come in," he said.
"So, I went in for my physical. And I was classified as 1Y, which means
you're not healthy enough for us to take you now, but we might take
you if things get worse."

And the bottom-line reason for the 1Y designation?

"It was because of my asthma." Mark reported. "The asthma that
had held me back from so many other things early in my life held me
back again. Only this time asthma kept me from going to war."

For the record, Mark would have served. But he did have a plan.

"My uncle had been in the military as a weatherman," Mark dis-
closed. "He spent all of World War II in the UK predicting weather all
over Europe. And he had convinced me that if I went in, I should put
in to be a weatherman."

All was not sunny for Mark, though. He'd really wanted those jobs
in Japan and Australia.

"It was good to be contributing to a larger effort, but the job itself
was not a good fit," he said. "So here I was. I was stuck in a job I hated.
I was living in New York City. And I hated living in New York City."

He stuck it out through two years and two assignments.

"My second year at Esso, they transferred me to the economics department," Mark recalled. "I'm in my mid-twenties at this point. My boss was sixty-four years old. And all we did was work on economic models designed to move money around to minimize Esso's tax burden . . . all done at a time when there were no big computer centers."

Crunching numbers and analyzing hypothetical financial data was not quite the professional path Mark had envisioned when he'd walked the stage at his MBA graduation ceremony.

"I hated that job. I hated it!" he said emphatically. "I just couldn't stand the sameness. And riding the subway to work every day . . . wearing a tie . . . I hated it!"

So, after this two-year stint, convinced he could not find professional happiness as either a ship traffic controller or a money manipulator, Mark decided he would move on from both Esso and New York City.

The question in his mind was what direction to take. Career-wise, should he look west to San Francisco or north to Boston, as both were hubs of what was fast becoming a technological revolution of never-before-seen proportions.

He'd dreamed of going to San Francisco, with a goal of toiling in what would very soon become known as Silicon Valley, so named because of the high volume of companies involved in the making of silicon-based semiconductors, the heart of modern technology. And had the Vietnam War not intervened, he'd have been there already.

But the driving force behind business growth on the West Coast was Stanford University. And with his undergraduate engineering degree from MIT, Mark decided that it would be easier to network and look for jobs in Boston, where MIT's reflection could be seen across the Charles River. So instead of San Francisco Bay and vineyards, he ultimately opted for Boston Harbor and snowy winters in a city he'd already come to know and enjoy.

Boston was hopping heading into the summer of 1969. The city was abuzz in the wake of the Boston Celtics stirring defeat of those dreaded Los Angeles Lakers in seven games in what was considered one of the biggest upsets in sports history.

Bostonians also swelled with colloquial pride knowing that many who'd received their educations in the area were heavily involved in what they saw on their tiny television sets that July, when Neil Armstrong took that "one small step for man, one giant leap for mankind."

Noteworthy as well was the unspoken tech rivalry between Boston and San Francisco. True, San Fran had its Stanford connection feeding into its soon-to-be-nicknamed Silicon Valley. But the Boston area had a feeder system that included MIT, Harvard, Boston University, Northeastern University, Worcester Polytechnic Institute, and even Boston College. Technology-based start-ups took root until the road around the perimeter of Greater Boston, Route 128, became known as "America's Technology Highway."

"I answered an ad, I think it was in the *Wall Street Journal,* for a job at a 'small' company called Digital Equipment Corporation," he laughed.

Digital Equipment, or DEC, was no joke. On the contrary, it was in the midst of a mercurial rise. Between 1960 and 1970, DEC would grow from a local computer company with 117 employees and revenues of $1.3 million into an industry redefining giant powered by some 5,800 workers and generating $135 million in revenue. Other organizations would eventually catch up to and ultimately conquer DEC, but not before it would reach its peak as a technology superpower in 1990, one that employed 120,000 people worldwide and took in a whopping $14 billion in annual revenue.

Mark was invited to DEC headquarters, a former wool mill in Maynard, Massachusetts, where he interviewed with the vice president of sales.

"At that point," recalled Mark, "they had never had a formal sales training program. And he offered me the job of starting that program."

That Mark had never actually held a sales position was not seen as a deterrent to either the VP of sales or his new employer.

"Admittedly, I had never sold. But what I had done was I had sold the concept of those internships while I was at Rochester in my first year. Then, in my second year, I had trained other students on how they could sell the concept," he said.

"So that was my experience," he continued. "And I felt I could do the job. So, I put together a four- to five-week course designed for new hires within DEC's sales force. They were hiring people all around the world at the time. So, they would bring them to headquarters in Maynard. I taught them the selling part of sales from a DEC perspective."

Excelling at his new position in his first 18 months, a pleased VP of Sales decided to offer Mark a healthy promotion to sales manager, which would have been comfortable enough given how many of the district and regional managers he'd already come to know. But Mark saw a speed bump on that highway.

"In those days, the route to advancement was that you got transferred around the world as you moved up the sales ladder," he remembered. "But I decided that I didn't want to become an international nomad for the next ten years. In a way, with my previous travels, I'd already done that."

But as Mark closed a window, by turning down the promotion, a door opened.

"I took a job at DEC helping launch the company's first minicomputer product line instead. We created a programming language called DIBOL. And we created a minicomputer business system that could be used by small businesses to run their companies," he said. "That was revolutionary in those days."

Again, Mark had proven his worth. He'd morphed through different assignments and was poised for advancement. So naturally, befitting his consistently unorthodox approach to his personal trajectory, it was time to find another challenge.

"I had become, basically, a prized designer and programmer. But again, climbing the ladder in DEC meant moving around the world. I didn't want to do that," he reiterated. "So, after six months of helping to organize this product line, it was time for me to move on."

The decision did not come without some critique from home.

"When I left Esso, my mother, for whom I had great respect, said to me, 'How can you leave the largest oil company of its kind in the world, and one of the largest companies in the United States . . . where you have a good job, you're earning a good living . . . and they like you?'" he recalled.

"This time, when I said I was leaving DEC she asked, 'How can you leave the largest minicomputer company in the world when it's doing so well and where you're earning a good living?'"

Mark didn't have a solid answer for his mother, the holder of a law degree from the Ivy League's Columbia University. But deep down he knew that at least part of the driving force behind his now seemingly nomadic approach to a career path had come from all the time he'd spent with his mother's mother—his grandmother. Simply put, though still not quite sure how he would navigate his way to the correct transition point, he knew he wanted to be a horse, and not a mule.

"So, I went out looking for a job as a sales manager or VP of sales at companies along Route 128 just outside Boston, which had a growing number of technology companies on it," said Mark. "I went calling on a lot of companies and, even though I didn't get hired as a VP of sales, because of my experience, I kept on getting offered consulting jobs. Soon I realized that what I had was a little consulting business. So, I decided to be a consultant."

Likely, Mark was having difficulty finding a full-time job because the country was in the midst of a mild recession. Undaunted, Mark began building what ultimately amounted to a pretty impressive client list. And that list gave him an idea. He would leverage his list to gain a full-time job.

"There was a good-sized consulting company in Concord, Massachusetts, named Billings and Reese," Mark recalled. "The company's principals had all gone to Harvard and Yale and Dartmouth . . . the Ivy League schools . . . so they were very formal. And here I was, this kid who came knocking on their door during the '71 recession saying, 'You need to hire me because I'll bring all this business in with me.' And I convinced them to do just that."

For the next three years, through his position at Billings and Reese, Mark consulted for some of the largest companies in the United States. He built a solid reputation for conducting extensive market research and creating product road maps. But the marriage was not designed for the long-term because Mark and his superiors had differing views for how consulting should be handled.

"We had a fundamental disagreement over how the job should be done," said Mark. "The modus operandi at Billings and Reese was that before you started the job you would go in with already developed answers. And you'd then work to prove these were the correct answers.

"I didn't like that approach," he laughed. "I liked taking on the job, digging in, and figuring out the best answers for a given situation on my own."

With those strong beliefs in a customized approach to client service clearly at odds with the tenets of the principals at Billings and Reese, the proverbial handwriting was on the walls.

"After about three years, we reached a point where we were at odds," he remembered. "And I believe it was the second time that I got called in and told to stop coming up with my own ideas that I decided it was time to leave."

And thus, came the inevitable call from his mother.

"She said, 'Mark, how can you leave *this* great job? You earn so much money! You have so many great clients! How can you do that?'" He recalled, "I told her 'I'll be fine' and I went ahead and started my own consulting company out of the basement of my house."

At that point, with a decent track record and an impressive resume given his relatively young age, cold calling for work wasn't as necessary. Work was finding its way to him. Yet something still ate at the former high school wrestler.

"I was grappling with this nagging feeling that was always in the back of my head," he said. "I was consulting for all these companies, and giving them what I thought were great product ideas. I'd do all this work for them. I'd tell them the right direction to go. I'd point out how to structure things. And then I'd watch them either not do anything or worse, totally blow it on the execution."

But one of those trips would prove fortuitous at a later date.

"I was thirty-four years old and working at a company called EG&G," recalled Larry Baxter. And things weren't not going so well. So, they called in some outside consultant named Mark Ain to look at some marketing options for us."

Mark performed his due diligence and then came back to report his findings and make some suggestions. At the end of that meeting, and with a common MIT experience to discuss, Mark asked Larry about doing lunch.

"At lunch, Mark told me my company was unlikely to do anything with his recommendations," recalled Larry. "And I found that funny. But in the end, he was 100 percent correct. The company did nothing."

The upside was that the two had developed a quick rapport. More lunches, a few dinners, and several games of squash later (Baxter conceded that Mark won most of the battles . . .) Mark had explained to Larry that, despite achieving a decent level of financial success, he'd grown frustrated with consulting.

So, Mark told his new friend, amid the contradictory factors of growing success and mounting frustration, he had reached a line of demarcation, resulting in a simultaneously easy and difficult decision.

"I'd finally reached that juncture," Mark recounted, "where I said to myself, 'I'm going to create my own company. I'm going to invent my own product, and start my own company. I'm going to show them all how to do it right!'"

And in Larry, Mark had found someone with whom he felt he could partner, a good guy who he enjoyed spending time with who happened to be an electronics genius.

Perfectly partnered, now all they needed was the white elephant— oh so rare and elusive—the niche somewhere out there in the business universe where the burgeoning technology of the day could fill an as-yet unfilled need.

Chapter Four

TAKING THE REINS/SADDLING UP

A horse is a horse, of course, of course.

—Jay Livingston (TV theme composer)

You'd expect the mantra would come from a source other than the composer and lyricist of the theme song for a 1960s television series about a man and his loquacious equestrian compadre.

Through the looking glass one might peer into the past for sage words of wisdom . . .

"Nearly every man who develops an idea works it up to the point where it looks impossible, and then he gets discouraged. That's not the place to become discouraged," Thomas Edison advised.

"All our dreams can come true if we have the courage to pursue them," Walt Disney opined.

"If I have a thousand ideas and only one turns out to be good, I am satisfied," Alfred Nobel shared.

Remarkable innovators all. But perhaps because, as the oldest of the five Ain siblings, Mark was most often charged with chauffeuring his grandmother from Brooklyn to his Long Island home and back for her frequent visits, her words had become firmly entrenched in his psyche.

"Be a horse, not a mule."

So, the concept of crunching numbers at the Esso Corporation, despite its place as one of the world's biggest and most powerful companies, held no allure for Mark. Nor did driving sales for DEC, even though the company was in the midst of redefining the role of computers in businesses large and small on a global scale. Even working for a very high-end and highly successful consulting company lacked long-term appeal, given an over-arching mandate to peddle the wares of the firm above providing true, customized consultation.

It wasn't about the money any more than it came down to pleasing his supremely accomplished mother, who consistently wondered why he felt the need to leave each of these situations just as he seemed poised for ascending their respective ranks. It was really about what he'd learned about himself and the role he knew he had to attain to truly achieve success, as he had come to define it.

"I was a mule," he confessed, "and I had to become a horse."

Ironically, the same was true for "Mr. Ed." To be successful in the path he'd chosen, show creator Arthur Lubin also had to learn about the value of horses.

Lubin, originally wanted to make a TV show about a talking mule. An accomplished movie director, Lubin had directed a series of films featuring the talents of Frances the talking mule.

Lubin wanted to bring Frances to the small screen but was not able to secure the rights. Sans mule, he improvised, securing instead the rights to a children's book called "The Talking Horse." Of course, of course!

Mark's eventual ride would, of course, be a horse of a different color. But still analogous with Lubin and Mr. Ed, Mark would need to adapt his game plan several times before finding the right ride.

"I had my own consulting business the next three years," Mark detailed. "But my aim there was not long term. I was using my extra time and effort looking for an idea around which to invent a product and start a company."

Mark had teamed up with fellow MIT alum and now good friend Larry Baxter, who brought to the table dual degrees in Mechanical and Electrical Engineering, and they began exploring ideas large and small.

To say they ran down every dark alley and kissed at least a few frogs in the process would be an understatement.

"At one point," Mark recalled, "we answered an ad in the Boston Globe. This guy said he had a fabulous idea, but we would have to come and meet with him in person before he would tell us anything about it."

Undaunted and at least a tad intrigued, the two aspiring entrepreneurs drove to a slightly seedy section in the city of Somerville, Massachusetts.

"We drove down one dark street after another until, on the darkest street of all, we found his house," recounted Mark. "But when we rang the bell the person who answered said, 'He lives in the basement around the back.'"

Sure enough, the gentleman with the earth-altering idea lived in the basement with his wife and three children.

"What have you got?" Mark remembered asking. "And he said, 'This is going to revolutionize the world, but you have to sign a confidentiality agreement in duplicate before I'll tell you about it.'"

"So, we signed his confidentiality agreement," said Mark.

And the idea that required this cloak and dagger level of secrecy??

"He had this idea for putting a radio inside a yo-yo," laughed Mark. "Imagine . . . you'd be able to play with your yo-yo and listen to the radio at the same time."

Yo-yos—and the world—kept spinning. Oddly, the radio-inside-the-yo-yo concept never took hold. Mark and Larry made a polite exit from that basement apartment and the search continued.

Not all unrealized ideas were without merit, it should be pointed out. But to run a realistic search for a next great thing required a sense for what could be accomplished.

"We actually had the idea for scanning barcodes like they do today in checkout lines," Mark confided, "but we decided that was too big an idea . . . we'd be up against too much competition."

"I came up with one idea I felt the world needed," said Larry. "It was a galvanometer laser deflector. So, we got into that a little bit until Mark said we should move on. But I firmly believe that had we stuck with that and worked hard we would have created a company that could have peaked at about a million dollars. That is, if we had put a million into it!"

"So, we drew up a list of 10 criteria for the kind of products we wanted to invent, and the kind of company we wanted to start," said Mark, "things like, the product could be sold for under $5,000 . . . that it could be sold to virtually any business in the world . . . things like that."

One thing, it turned out, would lead to another.

"We were actually building an inverter for small aircraft," said Mark, "because, in those days, there were no power supplies to meet the instrumentation needs of small aircraft. So, we were essentially building a power supply to meet that need, and that's when I discovered the time clock."

This connect-the-dots/eureka moment materialized thanks to yet another high school math team and MIT related connection—Donald Levy.

Levy, who had been Mark's collegiate roommate, had miraculously managed to virtually get his MIT degree in just three years. He'd then gone home to go into a family business where clocks were manufactured. The business would eventually go under, but not before an enterprising Donald, seeking a way to streamline costs at the company, had invented and written the code for a payroll processing product.

"When the family's company went out of business, Donald took what he'd developed and started a payroll servicing company like ADP," said Mark, referring to the human resource management giant.

Since Mark was simultaneously searching for his own idea and consulting, Levy invited him to take a look at his fledgling operation, which would come to be known as Manugistics.

"Donald called me up and asked if I'd take a look at what they were doing to see if there were ways to make the company grow faster," recalled Mark. "So, I went down, saw some obvious things they could do, and soon I found myself become a member of their board."

Shortly thereafter, Manugistics cut a deal to manage payrolls for customers of Chase Manhattan Bank. This required additional management, and Mark was drafted to work on building the teams needed to work with Chase customers on the migration.

"I came on board as a three- or four-day-a-week person to run the transition," Mark recalled. "So, I hired all the people who were going to go out and see the Chase customers and oversee all of these conversions."

This would constitute a sidestep from his search to find his own idea for a company, but Mark saw the move as a chance to both play a role in something dynamic within the rapidly evolving technological universe as well as a chance to work closely with good friend Levy. But in the final analysis, the move would become pivotal and precipitous within Mark's own trajectory.

"As part of what I was doing I met some people from an independent timekeeping company called Interboro Systems," said Mark. "It was a family business run by the Cohen Family . . . a father and his three sons. And they had this idea for automating the tallying of time cards."

"We were trying to create a clock that would record time," Bob Cohen, the youngest of the siblings would later recall. "But it was not working out the way we hoped. Somebody suggested we get in touch with Donald Levy. And Donald arrived with Mark Ain."

"Unfortunately, or perhaps fortunately," Mark said, "what they had tried to put together did not and, I think, never would have worked."

But there it was—*the* idea!

"Mark came back from New York quite excited," remembered Larry. "He said, 'I've got it!'" And I said, 'Okay . . . what is it *this* time?'"

"*This* time," the answer was computerized time clocks.

"I said, 'Why don't we start a company and we can invent that product?' and there it was," Mark smiled.

"It seemed, probably, more ambitious than what I had thought we'd start out with," said Larry. "But on the other hand, it looked as if nobody else was doing it. Either nobody was doing it or nobody had brought it to market."

The would-be horse had found a track to run on. Not the fast track, mind you. This race would not be a sprint. In fact, it would eventually become a global ultra-marathon. But, finally, Mark found himself in a viable starting gate.

It was about perseverance. It was about the stars aligning. And, ultimately it was, and would be, about time . . .

What it would not be, though, was without struggles.

Chapter Five

THE COMPANY YOU KEEP

You don't have to swing hard to hit a home run. If you got the timing, it'll go.

—Yogi Berra

It was December 31, 1976, and the crystal ball in New York City's Times Square dropped slowly to a landing that signaled the conclusion of the bicentennial celebration of the birth of the United States. The sulfuric smell of fireworks had seemingly filled the air all Leap Year. Elton John's "Philadelphia Freedom" had commandeered the airwaves. And a couple of guys named Steve Jobs and Steve Wozniak founded a company named after a piece of produce. For some reason, they thought the name Apple would bear fruit.

1977 brought change. Star Wars had us dreaming of a future in a galaxy far, far away . . . the first Space Shuttle piggybacked its way skyward atop a Boeing 747 . . . and Elvis signaled the end of an era as the King of Rock and Roll passed away on the throne in his Graceland master bath just in advance of the screening of Saturday Night Fever and the dawning of Disco Mania.

In late 1976, Mark, together with Larry Baxter and thanks to a connection via Donald Levy, had encountered a New York-based family named Cohen that ran a time clock company called Interboro Systems. The father and sons behind that company had come up with a potentially game-changing concept: they would create an electronic version of a time clock designed to record worker hours. Luckily, at least for the budding entrepreneurs, the Cohen concept held merit, but their efforts had thus far been in vain.

"Their product was a *kludge*," said Mark, using a term defined in the dictionary as a workaround or quick-and-dirty solution that is clumsy, inefficient, difficult to extend, and hard to maintain.

Mark's brash assessment was, of course, fairly accurate, as their version of the product in question simply did not work . . . at all.

"They were trying to take a mechanical clock and make it do what it couldn't," Mark opined. Still, Interboro had a concept that Mark, Larry, and Donald saw as an opportunity. Equally as important, Interboro had been successfully distributing mechanical clocks for several years, and therefore had a network of clients in and around the Northeast region of the United States. already using those manual clocks and services. A majority of those clients were savvy to the technological advances suddenly blossoming throughout all other aspects of the business world. So, they were primed to buy a better time measurement mousetrap if and when one became available.

"One of the big advantages of partnering with Interboro was they were one of the largest independent time-clock companies in the United States," said Mark. "And because of that, they really knew the time-clock industry. And they were able, in 1977 and 1978, to take me on a lot of sales calls where we called on people who wanted this new product . . . even though we hadn't actually conceived the product yet."

Why meet with prospects when you had no product?

"From these meetings, we were able to get a lot of good ideas. These companies would become customers, as long as we gave them the product they wanted," said Mark. "At one point early on, as I recall, we invited quite a few of their customers to a show in the Interboro offices in Manhattan. And we met the customers, got their ideas, and told them our ideas. It was very, very helpful having them as a partner."

Within Interboro, the brothers Cohen were in the process of taking the company reins from their father. They saw both that their attempt to modernize the clocks was bearing no fruit to that point, and that they had MIT-educated young guns willing to run with a concept they realized was beyond what they could do on their own.

At that point, the two groups decided they would form a company together.

"I remember flying to New York City to sign the documents with the brothers," Mark chuckled, "and they had their lawyer there. But, because a couple of them were very religious, they also had their rabbi there to bless the union."

Mark brought neither a yarmulke, nor a lawyer, with him that day. But it wasn't what he brought to New York that day—October 31, 1977—that was important. It was what he left with. He left with a company. Kronos was born.

"You know," Mark recalled, "Kronos wasn't anybody's first choice for a company name. But it was *everybody's* second choice. It wasn't even my first choice. I wanted Timekeeper. But we wanted to be sure everyone had a voice in choosing the company name. So, Timekeeper became the name that was attached to all our early Kronos products."

Of course, the bonus was that the company was now named after the mythological Greek God of Time (recorded through history as Kronos, Chronos, or Cronos), and for a young company whose primary strategy was to accurately record, and ultimately free up, time, the legend fit nicely.

"Obviously, we knew Kronos was the Greek God of Time," said Mark. "That's one of the reasons everybody liked it. So that worked out very well."

Coincidentally, what the fledgling company, now incorporated and armed with prospect input, required was exactly what the chosen name implied.

"Now . . . we were the ones who needed time," said Mark. "We had an idea. We had a company. And we knew what our potential customers wanted from us. But we needed to go from concept to creation. And that, in and of itself, represented a quantum leap. So, there we were . . . in a race against time . . . knowing that we had to be first to market . . . trying to make progress while we were working out of our homes . . . out of our basements and garages."

Working from home, but making steady progress. By the time the deal was signed and the name Kronos had been chosen, a prototype time clock had been assembled.

"We rolled our own computer," detailed Larry, who was not unfamiliar with building computers from scratch. "And we decided that 4K of RAM was more than enough for the life of the product . . . because who would possibly need more memory than that to run time clocks?"

Laughing later at just how little memory that would constitute in just a few short years, what Baxter would truly find funny later on was the mechanical engineering side of things.

"We made a box for the thing from an old printer," he recalled, "and we'd have preferred to go straight to all electronics, but we had to have physical time cards in the mix because potential customers said that's what their workforce was used to. So, let's just say that this was not a standard design project."

As Mark had evaluated Interboro's earlier effort, the resultant clock was likewise a *kludge*. But, in contrast to its predecessor, because the design was predicated on building an electronic clock from scratch versus adapting a manual clock to become electronic, this was an iteration with both promise and, ultimately, a patent.

"We built a prototype that could total people's hours as you put the time card in and pulled it out," said Mark. "And later we got a patent for reading and writing on time cards."

Knowing that there were other companies—some with deep pockets—exploring ways to bring technology to timekeeping, Mark knew that an accelerated push was needed. And that meant Kronos needed an infusion of capital.

"So, we incorporated the company on Halloween, October 31 of 1977," he said. "Then, that December, I went out to start raising money."

Perhaps as a sign that this would not be an easy process, Mark remembered his very first meeting to discuss a potential deal, which took place in Boston that December.

"I made my first attempt to secure funding when I met with a venture capital firm in Boston," he recalled. "But it was a cold winter's day and, after parking the car, I had an asthma attack while I was walking to their offices. My asthma used to bother me in the cold."

The money was not forthcoming that blustery day, nor for some time to come. But Mark was undaunted. Like the search for the right industry in which to create a company, the quest for money took extra time because the founder was very specific about his terms and conditions.

"We had a lot of people who were very interested in investing, but they were traditional venture capital people," said Mark. "And traditional venture capital people wanted either preferred stock or subordinated convertible stock. And I said, 'If I'm going to be working full-time at this company, you're going to have the same common stock that I have. Otherwise, I'm not doing your deal.' And that was pretty much a nonstarter for the traditional venture capital firms."

That he held to his guns would prove critical for both the short and longer term for the fledgling Kronos.

"Had any of those venture capital firms invested," he said, "they probably would have fired me during the '81 recession and the company would have never grown. So, it wasn't hard to raise money. It was hard to raise money on my terms, which was the issuance of common stock."

Undeterred, and with a clear vision of how he wanted things structured financially, Mark doggedly set up meeting after meeting, consistently garnering attention but unable to proceed on the terms he felt would be best for him and the future of Kronos. Then, after those six months, pay dirt came in his new high-tech venture via an old-fashioned pipeline: he was introduced to an investor, who happened to be the co-director of a charity with Mark's father Jack.

The date stands out as a milestone amid milestones in Mark's head. It was June 5, 1978. The investor, a New York–based venture capitalist named Tony Parkinson, who at the time owned, among other things, the National Hockey League's New York Islanders, agreed to a common stock deal, one that ensured Mark would remain at the helm.

"With that money, we rented a small space on the third floor of an ironworks plant in Boston, Massachusetts, to complete our product," he recounted.

Out of their respective garages and basements, the team now had a dedicated work space to share and propel things forward. Sure, it was a third-floor space. And iron was being smelted and fabricated down below. But every company needs to start somewhere. In this instance it would be amid the stench and heat of burning iron. Kronos had a lease on its very first world headquarters.

Chapter Six

WHEN THE TIME WAS RIGHT

Success is not final; failure is not fatal: it is the courage to continue that counts.

—Winston Churchill

The year 1979 brought calamity and change.

The burgeoning nuclear industry took a hit when a failed cooling mechanism at Three Mile Island caused a partial meltdown in Reactor #2. Young couples in China were told that as part of their government's population planning policy that they would only be allowed one child. And sixty-six Americans were taken hostage in Iran.

It wasn't all bad news. Amid a backdrop of The Sugerhill Gang's "Rapper's Delight" and Rod Stewart's "Do Ya Think I'm Sexy?" Americans flocked to movie theaters to see Martin Sheen surface in "Apocalypse Now" and Mel Gibson quietly take on a collection of crazy post-apocalyptic hooligans in "Mad Max." And going downhill became a good thing with the introduction of the snowboard.

It snowed in the Sahara Desert for a grand total of 30 minutes. A tsunami killed 23 people in France. And Great Britain elected Margaret Thatcher as Prime Minister.

Along the way, two University of Connecticut basketball fans, unable to attend all their favorite·team's games, licensed space on a new concept called cable television, giving rise to an organization called ESPN. Canadian board game Trivial Pursuit rewarded those with a command of inconsequential knowledge. And music got mobile as Sony introduced the first Walkman.

Meanwhile, at Kronos World Headquarters, things were not necessarily going well, but they were going. Finally emerging from their respective basements and garages, the dozen or so employees of the fledgling organization found themselves toiling for the first time under a singular roof. True, conditions were less than ideal. But the fact that they could interact, and test, and postulate the possibilities of actually creating a breakthrough product, put wind in their sails.

"Finally getting everyone under one roof really got us on track," recalled Mark. "Thanks to Donald Levy we had connections and a partnership within the industry. And thanks to Larry Baxter, we were on our way to having an actual product."

Mark knew the demand was there for a better, electronically based way to capture time data. His constant trips to both establish a prospect pipeline and seek input for the design phase proved that out time and again. Many would, and actually did, question why he was so dogged in his pursuit of those aspects versus staying put in the office and working on the prototypes. But Mark had an ace in the hole in the form of Larry Baxter.

"I couldn't do what Larry was capable of doing," Mark confessed. "Even at a place like MIT he was considered brilliant. Think about that!"

"Mark is too self-effacing," Baxter would later say. "He was the great organizer . . . the way he brought the team together and kept everyone motivated. Without him there would never have been a Kronos."

True, Mark's undergraduate degree looked like the others. But while the rest would grow in their knowledge and stature as engineers, that Rochester MBA experience had Mark ready for the driver's seat. And he knew that in Baxter, who would become recognized as one of the world's absolute experts in the realm of capacitive sensing, he had a partner capable of delivering on the promises he was making in the field. The trick was to gain momentum and keep it.

"I was full guns," Mark said. "I wanted to be in there seven days a week. And there were very few days when I would pop in off hours and find myself as the only one there.

"One thing I instituted right away," Mark continued, "was that every single Monday we would hold a staff meeting. And at those meetings, because there were so few of us at the time, each person had to give an update on what they were doing, what progress they had made, weekly goals, and weekly accountability."

Because of this accountability, Mark would often find Kronos employees, like himself, in the office working on weekends in an effort to achieve their weekly goals.

There would be no overnight success. But the stakes were perhaps higher for prospects than even the intrepid Kronos development team realized. Because Kronos' first customers would be using this newly invented piece of technology to automatically record labor hours, knowing full well that their employees represented the most important, and costly, aspect of their respective organizations.

Finally, through perseverance, ingenuity, and a dose of technological brilliance, a prototype was fashioned—not a perfect, finished, fine-
tuned masterpiece, but one that mainly performed the appointed task.

"We never expected to simply build a clock and then be done,"
said Mark. "In fact, throughout the history of Kronos, our motto has
always been 'If it ain't broke, fix it anyway.' And it's a philosophy that
has kept us moving forward and allowed us to keep our customers
happy for a pretty long time."

Experimentation continued as Mark hit the road again in search
of additional funding, as the possibility existed that the company could
run out of money before a workable clock could be developed. But
suddenly there it was: the Timekeeper!

"Those early clocks," laughed Baxter. "They had their issues. But
we ultimately built a prototype of a time clock that could actually total
people's hours as you put the time card in and pulled it out."

"And," chimed in Mark, "we got a patent for the process of reading
and writing on time cards."

The design team, with Larry Baxter on point, had utilized this
newly emerging microprocessor technology. The resultant "clock" was
essentially a computer that totaled time cards, applied work rules, automatically reported time worked to the payroll system, and eliminated
errors. In terms of labor management, particularly where it came to
the front lines, this was a seismic technological leap.

Thus, and thanks to Donald Levy and partner Interboro's client
base, in late 1979, Mark and the team carefully loaded a couple of
newly minted Kronos Timekeepers into an aging Interboro hatchback
bound for New York City, where new Kronos employee Aron Ain, the
youngest of Mark's brothers, would assist the Cohen team with the
installs.

Mark had convinced Aron to come on board upon his graduation
from Hamilton College in New York.

"We joke now that he started out sweeping the floors and cleaning
the bathrooms in that first work space," said Mark. "But he was very
good at installing the clocks and later he was likewise great at selling
them."

According to Kronos legend, the very first clock, for lack of ade-
quate wall space, was hung in the employee restroom of a copying
firm near Times Square, which likely made for some interesting early-
morning punch-ins.

"It wasn't glamorous," laughed Mark, speaking more about all the
early clocks than about the individual restroom placement. Measuring

time wasn't meant to be glamorous. It was meant to be efficient, accurate, and dependable."

Problem was that these early clocks had all three of those characteristics . . . just not all the time.

Of particular note, though not a prevalent issue, was the fact that on a rare occasion, but without any foreshadowing, the microprocessor would stop working at the precise instant the high-speed clamp had closed on a time card. Because the clamp was equipped with an electromagnet designed to close the clamp rapidly, when it jammed it heated up very quickly around the cardboard time card, causing the unit to burn out in a puff of smoke.

"One time early on," Mark recounted, "Aron had just installed a clock, and apparently the clock had malfunctioned. So, I got a call from the new customer on Long Island and he said to me, 'I've got an axe in my hand, and I'm standing next to your brother and your time clock. Which one should I use the axe on?'"

The team was perplexed—and concerned. Should this proverbial crack in the dam start to spread, and word got out to prospects and investors that this new technology was, in fact, a fire hazard waiting to happen, the whole dream could come tumbling down. Kronos could be done before it ever got fully out of the starting gate.

"We had to be accurate. And we had to be reliable," Donald Levy would later say.

"The question," Larry Baxter would remember, "was whether it was a hardware problem or a software problem."

The issue was perplexing. How to fix a problem when you couldn't isolate where the problem was coming from?

Enter Larry Krakauer . . .

Another MIT alum, Krakauer's resume would prove to be the much-needed needle in a haystack. In fact, the fit for the immediate need would likely come from even longer odds . . . a needle hidden inside a virtual mountain of hay.

Larry had received his PhD in electrical engineering from MIT in 1970. From there, he'd become a rare breed as, for the ensuing five years, he directed software development for a small start-up. Then, he'd changed companies and job functions as he turned his talents to hardware development for Codex.

"Yes," he'd recall in his own blog many years later, "Remember, I'm an electrical engineer. But I switched from managing software to digital hardware design!"

Having achieved success within multiple disciplines, but having also started a family in his early thirties, Krakauer had already made up his mind that he wanted to be part of an entrepreneurial venture, but one with very specific parameters—entrepreneurial, but not a start-up—a place where he could have a stake in the company, but not be a founder logging hours the way someone like Mark had been doing. He was aware he was being purposely picky, but parenthood had adjusted his priorities.

"I made this trade-off in the direction of family, and decided not to get involved in a raw start-up," he'd recall. The search for that hybridized fit had borne no fruit, until he walked into the MIT Alumni Placement Office and saw a letter tacked to the office's bulletin board written by a fellow grad named Mark Ain.

"Kronos' innovation," Krakauer would recall, "was to apply the newly emerging microprocessor technology to essentially put a computer inside the clock, allowing it to apply work rules consistently and fairly, eliminate errors, total the time cards, and automatically report the time worked to the payroll system."

Krakauer was intrigued. Kronos was just beyond the "garage" stage. And they had a modicum of traction as they were by that point shipping product. Most intriguing to him, was they had a problem that these other quite brilliant ex-MITers were having trouble isolating.

"Kronos went looking for someone who could . . . in the short term . . . resolve the problem of the smoking solenoid," he said. "They needed a talented engineer familiar with both hardware and software. They needed . . . me!"

After what Krakauer would facetiously deem "an extensive interview process," meaning his resume, seemingly alone, fit the bill, he was offered the job of Chief Engineer at Kronos.

"Finding Larry Krakauer . . . or more accurately Larry Krakauer finding the letter I posted at MIT . . . was a pivotal turning point for Kronos," recalled Mark. "As I said, our platform was built on being accurate all the time."

"It had turned out that the product was not as ready to sell as Kronos had thought," Krakauer said, "although the misrepresentation was not deliberate."

Larry set to work. From the outside looking in, this immediate problem appeared to be a hardware issue, but his dual expertise quickly allowed him to ascertain that the hardware was indeed working as prescribed. So, it was time to track his way through the software.

Once I started reading the program," Krakauer recalled, "it took only a short while to trace the intermittent lockup problem to the software, to a very badly designed interrupt system."

Krakauer and his "team" of two other people were then forced to almost completely rewrite the clock software, thereby, again quoting Krakauer, "saving Kronos' corporate ass, if corporations can be said to have an ass."

The initial issue was solved, along with several others that the core team was not yet quite aware of, and Kronos had discovered a problem solver who would help propel the fledgling organization forward. Larry Krakauer was summarily promoted to vice president of engineering, a position he would hold for almost twenty-five years.

The ensuing and evolving engineering team would allow Mark to begin what would become an ongoing quest to not only fix what was and what wasn't broken, but to innovate ahead of an ever-evolving technology curve. By committing to constant innovation, he would not only transform the timekeeping universe, but also the world of what would one day come to be called workforce management.

But, as he had learned in his personal quest to find his way to the top as a business leader, the foundational key lay in finding the right people to lead.

Chapter Seven

TEAMING UP FOR THE LONG HAUL

*Teamwork is the ability to work together toward a common vision.
The ability to direct individual accomplishments toward organizational objectives. It is the fuel that allows common people to attain
uncommon results.*

—Andrew Carnegie

The early 1980s was a whirlwind. California Governor Ronald
Reagan, a former actor who'd made his name playing opposite
chimpanzees in movies, was president of the United States. PAC-MAN,
an arcade game sans pinballs, was a digital rage. Americans began
screaming, for the first time, that they wanted their MTV. And *Thriller*,
Michael Jackson's follow-up to his solo *Off the Wall* album, forced
the music industry to invent a new best-selling category: Diamond
Platinum.

Underneath it all, and unseen under the hood of all things electronic, microprocessors were reshaping the present—and the future.
Sony had people dancing in the streets with its burgeoning array of
mobile sound systems. IBM brought computers home with its first personal computer, or PC. And computer systems began interacting via
phone lines over something that would eventually be called the Internet.

Meanwhile, Mark and his merry men (and women) were forging
ahead in the old ironworks shop in Cambridge. It wasn't glamorous.
And a profit had yet to be made. But with a product shipping and
mostly performing as advertised, that elusive glimmer of hope for success was growing brighter. Word was spreading about a little company
that had indeed built a better timekeeping mousetrap. As had been
the bottom-line objective when Larry Baxter and Mark had evaluated
everything up to and including yo-yos with built in transistor radios,
Kronos had been founded within a relatively small industry—time
clocks and the keeping of workers' time—where the technology was,
putting it bluntly and pun intended, behind the times, as the basic
time clock then used hadn't dramatically evolved in over 100 years.

By melding the emerging power of the microprocessor with the need for a more reliable and efficient way to collect time worked, all within plain sight of the gigantic companies that dominated the landscape at the time, they had set the time clock in sync with the rest of a rapidly evolving way of doing business. They had taken a first step in ultimately changing the way labor was managed.

There was no meteoric rise to the top, however. To the contrary, Kronos was better represented as *The Little Engine That Could*, grinding its gears on a slow climb up a steep incline, making baby steps of headway, building and improving one clock at a time, teetering occasionally toward disaster, yet seemingly finding its way at precisely the right time.

The journey wasn't for the squeamish. Truth be told, there was some noticeable attrition in the earliest of years. An early partner bolted. A few employees saw other avenues that they deemed more promising. Others didn't work out. But Mark, emboldened to finally have a viable product, one he knew could be improved and enhanced to solve more complicated customer needs, saw a bigger landscape forming. He'd seen it before at the young Digital Equipment. Intelligent expansion would be important. Finding the right leaders would be key. Defining and refining his own role, and having enough trust in the capabilities of others to uphold their respective ends of the bargain, though not always Mark's strongest suit, was paramount!

In Larry Krakauer, he'd found Waldo, the lone blended engineer in a veritable sea of technology specialists who was capable of handling hardware and software issues with equal expertise and tenacity. Fortuitously, Kronos was likewise Krakauer's eye of the needle, as the company had moved just north of start-up, seemed to have legs, and truly needed his hybridized background. And the timing of his arrival and subsequent decision to stay was amplified by the fact that while Baxter had relished the early quest to develop the product and launch the company, he yearned for other challenges outside the realm of workforce management, and did not wish to manage engineering.

All of this . . . the product development, the shipping of time clocks, the hiring of Krakauer, and the general expansion of the business . . . happened because Mark, who as a natural decision maker was never singular of focus when it came to his fledgling company, respected those who held expertise that he lacked.

One of those critical experts for Kronos had arrived just ahead of Krakauer. Laurel Giarrusso would become Kronos CFO but had been recruited by Mark to come on board part-time to stitch together the finances and internally audit how money was being spent.

Laurel had started her professional career at Arthur Andersen, and had recently taken a CFO slot at a small tech firm that was an Andersen client, a move driven by her realization that she enjoyed the excitement and general vibe within many of these technologically driven start-ups that were sprouting up like dandelions on an untreated lawn.

"Mark appeared on the scene to recruit a marketing executive for our company," Laurel recalled. "What struck me right away was he was doing this type of recruiting while he was still trying to figure out what business to start. I was really impressed by the fact that he didn't focus his passion around a particular product idea. Rather he wanted to enter an industry that he described as being led by large, lumbering giants that couldn't respond to a competitive threat from a company that could move quickly and change the whole paradigm of the industry through technology. I just thought that was brilliant!

"Quite frankly," Laurel continued, "I think one of the reasons Kronos did so well was that single decision to focus on an industry like that, because it gave us the time to really get our feet under us and learn from doing. You get to know a product line and a company at that same time. That idea, when I was talking to Mark way back then, it just really impressed me. *He* really impressed me.

"I always considered Mark to be a very calculated risk taker," Laurel assessed. "He wasn't the kind of risk-taker who would just throw everything out and start from scratch. Rather, he wanted to take the time to think about what he was going to do before he got into it. I liked the fact that he was a generalist and he certainly had the ability to sell the market. He certainly had the ability to sell products. But he could also sell people, and he could very effectively sell ideas. So, when Mark came to me and said we really need someone to come in on the financial side . . . that was very attractive to me."

Mark had found an id to his superego in terms of handling the day-to-day financial operations for Kronos. In a way, he was breaking a business barrier in that Laurel's rise to C-level executive was not yet near the norm for women. Yet Mark's choice was not based on gender. It was based on talent and personality.

"I respected Laurel," said Mark. "She was a straight shooter and called things by the numbers. We needed someone like her."

And so, it came to be that Laurel was uniquely positioned to step forward and tell the boss that as the company was changing, he, too, had to change.

"Kronos was just getting ready to go from an R&D company to a company selling product," recounted Laurel. "Things were happening. And Mark was excited because we were finally going into the sales stage. We finally had a product and we were going to get to sell it. And Mark knew he could sell anything! So, he was out there going to customers, and getting sales . . .

"But," Laurel continued, "I knew that we needed to do things like raise money. We needed to be thinking about the future and where we went after this initial sales push."

So, one day Laurel found Mark as he packed his now well-traveled demo time clock for his next sales trip and sat him down."

"I recall that meeting well," chuckled Mark. "I remember that Laurel put it very gently. Basically, she sat me down and said, 'Mark, do you want to be vice president of sales? Or do you want to be president of this company? If you want to be president then you have to stop traveling every week and start raising money.'"

"In a business environment," stated Laurel, "I didn't mind being the bad cop. And Mark needed someone to play that role with others he'd recruited to join the company. But it was also true that it had gotten to the point where the sales and marketing work that Mark was doing were getting in the way of doing those other critically important things that were important to Kronos so that we could grow and thrive."

Truth be told, Kronos was generating revenue. But the ramp-up costs—salaries, sales trips, rent, materials—had eaten away at that initial funding base.

Mark thought he already had an ace in the hole. Tony Parkinson, his initial major investor and someone who seemingly had very big purse strings, had been impressed by the progress and in-roads Kronos had made. Listed as a founding member of the company and a board of director, Parkinson had deep knowledge and belief in the company's potential. And he'd already intimated to Mark that, given how things had gone, he would be willing to spearhead a second round of funding.

"I knew Laurel was right," Mark admitted. "Kronos was not going to survive without another round of funding."

"But," Mark continued, "Tony had already said not to be worried. He'd have another round available when we were ready."

But Mark was soon to learn a valuable lesson that sometimes timing and market forces are driven by something other than money. As he readied his pitch for Tony, Mark picked up a copy of the *New York Post*, the gossipy predecessor of *People* magazine and later *TMZ*, and learned that Tony, well known in New York circles, was in the process of a divorce. Thus, his assets were frozen and unavailable.

Financial safety net disengaged, Mark would need to step away from sales, at least in terms of weekly trips and cold calls, and take on that full-time role as the president of a company very much in need of private equity capital.

Problem was, if he were out chasing investment capital, still with an eye toward not ceding away his control of the company, who could step in and continue to establish territory and generate sales income? Most of the team was focused on operations in one capacity or another.

But there was one viable option. Not too far back, Mark had hired a young man fresh out of college. The young fella had been unsure what he wanted to do in terms of starting a career, but Mark saw potential, had him interviewed by some other partners, hired him, and immediately handed him a mop and bucket.

But Aron Ain, the youngest of Mark's brothers, was undaunted. Brilliant and gregarious, but not at all averse to doing the dirty work, Aron was a quick learner. Soon he was in charge of picking up the lunchtime sandwich order. Not long after that, he was learning how to install and troubleshoot the clocks.

Aron also had a knack for thinking outside the box where sales were concerned. At the time, the focus was less on sales to individual companies and much more so on signing distributors to act as sales agents, replicating the agreement Kronos had with Interboro and the Cohen Brothers in the Northeast. And it was Aron who came up with an innovative marketing idea to attract resellers. In an age before information could be looked up via computer in the home, Aron took to the Brookline Public Library in Massachusetts, armed with a pad of paper and a pen.

"I don't remember what led me to the library," Aron would later recall. "We kind of invented our game plan as we went along. And the library had a whole stack of Yellow Pages from all over the country. Seems like a funny thing for them to have. But that was before computers. So, I could sit there and flip through the Yellow Pages for places like Charlotte or Phoenix and look up . . . the big resellers of time clocks.

"Took me a whole day in the stacks writing down all these names and addresses," said Aron. "Then I wrote a custom letter for each place, put their name on it, put in a brochure and mailed it with a card saying, 'Thank you for your interest in Kronos.'"

He'd found resellers throughout the country during this research mission. And though the follow-ups could sometimes be awkward, they were also quite fruitful.

"None of them had expressed any interest in our product," recalled Aron. "And on a few occasions, I didn't need to follow up . . . I got phone calls from some of these guys who called to ask 'How did you get my name?' And I just played dumb. I said I don't know, but now that I have you on the phone . . . even though you don't think you asked about it, could I talk to you? So, I told them about our product without mentioning I'd found their name at the Brookline Library.

"Quite a few of those connections," said Aron, "became resellers for Kronos."

The company jack-of-all-trades, Aron, who had majored in economics and government in college, had a hand in almost everything at the company with one exception.

"It was a small start-up company that was pretty loosey-goosey. So I just pretty much did everything I was asked to do, from installing clocks, supporting clocks, doing marketing, carrying boxes, being the first call on the alarm list when the alarm went off at 2 a.m.," he recalled, "and yes . . . I may have cleaned up a bathroom or two . . . "

Mark would still have preferred to be headed for Logan Airport. But, with frank guidance from Laurel, he knew where his priorities needed to be focused.

"I looked around the company and quickly realized that the only person I could send out was Aron," Mark said, "He'd been with us for about a year at that point. He'd been out to install and restore units. He was heading up telephone support for the dealers. He knew the product. It wasn't something I had to think about for a long time . . . he was the only alternative."

Mark would dedicate a majority of his time to securing a next level of funding. Aron would board a plane for the Carolinas. And the expanding organization would subtly evolve foundationally. Aron would continue a step-by-step rise that would one day find him running the company, but not before Mark would officially take the wheel and guide Kronos to near-unprecedented multi-decade success in the world of business and technology.

"In the three months it took to raise that money, a lot changed," said Mark. "One of those was that, in continuing to trust my instincts where finding the right people and placing them in the right positions was concerned, I moved into my main role as Kronos President and CEO."

Chapter Eight

WIND IN THE SALES

Build a better mousetrap, and the world will beat a path to your door.

—Ralph Waldo Emerson

In 1899, a man named John Mast of Lititz, Pennsylvania, filed for a patent for the very mouse trap that immediately springs to mind virtually every time the phrase "a better mousetrap" is uttered. A heavy spring wire swings down when an unknowing rodent nibbles on a cheesy bait (or sometimes peanut butter), breaking the mouse's neck.

It was not the first invention designed to rid homeowners of the pesky critters. But for well over a hundred years, it has remained the gold standard, in spite of the fact that the United States has issued over 4,400 patents for "better mousetraps" in the ensuing years.

The question of whether any of those mousetraps were actually "better" is almost moot. Likely there were several iterations that were better at one aspect or another where it came to attracting, tricking, and dispatching of mice. But in those last nearly 120 years, only about 20 of those 4,400 ever made money. None displaced Mast's design or success.

In 1980, Kronos was not trying to rid companies of pests. Rather, where traditional time clock companies were concerned, Kronos *was* the pest because, at least metaphorically, the upstart company had built a proverbial better mousetrap with regard to the collection and output of employee time.

Not that Kronos was alone. The advent of the microprocessor, in a decade that would serve as a precursor to the dawning of the Internet era, was in the process of redefining processes and perceptions about the way people and companies performed. Mark Ain and his early compatriots had seen this transformation coming, and had defined an untapped niche within the realm of workforce management where they felt they could use technology to affect a paradigm shift—timekeeping.

With a version of the Kronos Timekeeper clock up and ready for deployment, Ain's young company still faced an uphill battle—one that likely felled many an aspiring mouse-catcher before them. A sizable chasm lay between building a better mousetrap and selling it into profitability.

Lacking a full-time sales force of his own, Mark had decided the Interboro third-party reseller model was the best way to gain traction in the marketplace. So, a majority of his early out-of-town trips were to meet with resellers in an effort to convince them that repping this new Kronos technology would reap benefits for both parties. Those trips were more laser-focused after brother Aron's research began to yield bona fide prequalified leads. And Mark conceived a pricing structure that was designed to incentivize resellers even further.

"We began by allowing exclusivity by geographic region, so resellers were not competing against each other for business," detailed Mark. "But in return for exclusivity, and to continue as part of the network we were putting together, they had to agree to a quota."

A quota system was not new in the reseller market. In fact, for most higher priced commodities of the day, a quota was more the norm than the exception. But in an effort to have salespeople push harder for Kronos than any other organization, Mark came up with an innovative pricing structure.

"The price they paid for our products was determined by what percentage of quota they achieved for a given quarter," Mark revealed. "That was a very revolutionary concept at the time because the discount didn't apply to those units sold over quota. As you went over quota, the discount applied across the board, to all the units sold that quarter."

The discount was not static, either. The more units sold over quota, the bigger the discount and therefore, the bigger the margin to the reseller.

"People laughed at me when I outlined the offer," said Mark. "Not the resellers, mind you. They were very receptive. The product demoed well. And there was a market. But with the advancing discount . . . it worked extremely well."

Momentum was building. A sales network was being constructed. Interest was rising. The handful of units Interboro had sold at the end of 1979 had been installed, along with others within a cab ride of Kronos' Boston office space, and a few referenceable customers were being won over. Yet sales were not exactly categorized as smooth sailing.

"People didn't trust computers," Interboro's Bob Cohen remembered, "not at the time, anyway. And we were selling a $5,000 solution to a $500 problem. Because that was the difference in price. The standard manual time clock retailed for about $500. And the Kronos clock went for five grand!

"On top of that, you had to be part magician to sell it," Cohen continued as he detailed the demo process. "At the end of a given time period, the clock would create a summary, which it would duplicate on a card. And 60 percent of the time . . . perfect! The other 40 percent, not so much."

So, Cohen and his cohorts used a bit of sleight of hand in those early demos.

"If the numbers were wrong, we'd show it very quickly and then put it aside," he chuckled. "And we tried to keep a pre-done card nearby."

With a burgeoning and increasingly expert engineering team working out the bugs, the need for tiny trickery ended and the reseller search intensified. This was the stretch where Mark, who was both an excellent salesperson and the company's best demo person, found himself flying to multiple cities every week to convince resellers that Kronos was the future of timekeeping.

"At the beginning of just about every week, I'd get on a plane with the timekeeping device as my carry-on," Mark said, "which was probably easier than it would be today because the clocks were bulky . . . they weighed about forty pounds . . . and I think airport security today would have a fit!"

Frequent flying had its rewards. The TimeKeeper time clock was indeed a better mousetrap. But because of the huge price differential, any and all leaps of faiths required seeing in order to create believing in the value a purchase would bring. Soon Mark would need to focus more on leading the company. But during this stretch, selling Kronos to resellers was his primary focus.

"The technology the clocks delivered meant that you didn't need a roomful of people tabulating workers' time," said Mark. "But there was a leap of faith needed because computers were nowhere near a part of people's everyday lives at that time."

"I had one customer who complained to me that they weren't seeing any cost savings after putting clocks in place," Cohen laughed. "So, I went out to see what was wrong. And what was wrong was that they had the clocks tabulating and reporting time . . . and a roomful of clerks retabulating the time to make sure the clocks weren't making any mistakes!"

"That was the case on more than one occasion," confirmed Mark.

To prove the value of selling this new technology—these marvelous examples of harnessing the power of employing microprocessors into the arena of timekeeping with labor management—Mark would regularly travel to a new region for Kronos, meet with a potential reseller, and then head out with that reseller on a sales call, where he would handle the demo, and hopefully make a deal that showed just how marketable the clocks were.

Most trips resulted in hits. But some were more memorable than others.

"I'd set up a meeting with a potential distributor in Denver," Mark remembered. "He'd set us up for a couple of sales calls and the first one was at a Pepsi bottling plant. We arrived and were shown into a boardroom and the fellow we were pitching to came in and asked if we'd like anything to drink. Without thinking I said, 'Sure . . . I'll have a Coke.' Dead silence. I said, 'I guess I'll make that Pepsi from now on.' But we ended up selling all of their plants. They became our first customer in the Denver territory."

Most trips worked out. Mark arrived with a time clock in hand and a message in mind.

"We would save them time by properly recording the time their employees worked," he stated. "They would save on the hours it took to record and tabulate time. They would also save on not having human errors in the equation."

Human errors meant when someone wasn't paid enough you had to go back and fix it, and that took time. It also meant that if you over-paid someone, they didn't usually complain, so you didn't get that money back. And that was the answer to the $5,000 versus $500 for a time clock conundrum, because correctly collecting hours and paying people the first time equated to big savings at both ends of the time and attendance equation.

The trick was getting the word out there—to get to a point where inbound calls were coming in, seeking to purchase the clocks. That meant acquiring referenceable users of the clocks. Mark knew this. So beyond simply demonstrating the way the clocks worked, he also emphasized the upsides of becoming an early adopter and reference-able customer as part of his early sales pitches to recognizable companies like his new favorite soft drink.

"Pepsi, as I said, became our first customer in the territory," said Mark. "I explained to them that if they were the first customer in the territory to put our time clocks in, and if they'd then agree to be a reference for us, I'd personally give them a great price.

"And secondly, I told them," Mark added, "because you're going to be a reference, we're going to kill ourselves to make sure everything works just the way it's supposed to work. Like in other regions, this was a very effective pitch. I would say our batting average when making these arguments, along with the potential savings the clocks presented, was at or above 80 percent."

That didn't mean all large orders were discounted. Mark also recalled a meeting in a boardroom full of executives representing a large Midwestern retail chain. With a big order on the line, and a salesperson who had already promised a discount, Mark took an opposite tactic.

"I told them I couldn't offer a discount for one very good reason," said Mark. "I explained that, with the size of their order, discounting would put a financial strain on Kronos. And I didn't want to only offer them time clocks one time. I wanted them to be a long-term customer. And the only way for us to continue to make our offerings better and better was for us to make enough money to reinvest in constantly improving the product."

And?

"And we got that order, too . . . at full price," said Mark. "Even though our front-end sales guy was a little blindsided, that was a very good day."

Landing good-sized orders was important. And doing so the way Mark was able to do it was certainly impressive to prospective dealers. But as the head of Kronos, the former high school wrestler was also getting in close and making tough decisions.

"I was very careful to make sure it wasn't just about securing a distributor," he recalled. "It had to be about securing the right distributor. I had to like them. I had to believe that they were sold on the product, and that they would have Kronos' interests at heart. Then, if they checked all the boxes, I would offer them a contract."

But Mark was also wise enough to know that a fully outsourced distribution channel by region was not necessarily the best way to construct a sales network. It was expedient. And it was effective. But simultaneously taking a short- and longer-term approach to sales would allow Kronos to grow right away, while also keeping an eye on a horizon Mark had set for his company.

"We hadn't made any money yet," he said. "In fact, as we went through the necessary growth cycle in terms of refining and bringing the product to market, we were actually losing money. But I believed in what we were doing. And I set my sights on Kronos becoming a $100 million company."

Kronos had already inked deals with a few distributorships that had been selling Cincinnati manual time clocks. They were good choices because they already had a foothold in the industry. And, like Simplex, they represented a corporate behemoth that somehow didn't make the synaptic leap required to see that the advent of the microprocessor could redefine the landscape they ruled. Or, if they did, Mark saw no evidence of that for the time being.

But not all Cincinnati distributors were created equal. Some were not worth signing for a myriad of reasons, be that a lack of interest on their part, a lack of resources, or an inability to pass the Mark Ain sniff test, meaning that at the end of the day (and the visit), Mark didn't have a good feeling about them. And, it would be fair to say on the flip side, there were likely some distributors who either decided Kronos and its technology were a fast-passing fad, or they didn't like Mark either.

Rather than focus on the negative side when a trip didn't necessarily yield a grand slam, Mark remained content to chip and charge.

"Something good came from almost every trip," he said. "We didn't close every sale or sign every distributor. But sometimes, when I met someone who was the right fit, I would sign them up as a direct employee."

That concept bore fruit on an early trip to Chicago, clearly one of the bigger markets in the United States. Mark had hired a former Simplex salesperson to "drive all around the Midwest following up on leads" generated back in Massachusetts. Mark flew out to join him and the duo closed three sales in just two days, including a referenceable one with a McDonald's franchisee. It was then that Mark decided that he would try to establish a full-fledged sales office, a first for Kronos, in the Windy City. The reason, like his ongoing research into how to constantly improve the clocks based on customer input and feedback, existed on a couple of levels.

"We wanted sales in a major market. That was for sure," he said of the concept. "But we also wanted a direct sales office so we could calibrate how the deals were being made . . . what it was that generated interest and, ultimately, what it was that would turn a prospect into a sale."

Wind in the sales! But there were counter-winds starting to pick up—winds that would breed yet another change in Kronos' direction.

Mark prided himself by then at being the type of leader that recognized the necessity to assemble a great team and then listen to the
experts he'd brought into his circle. And not long after his return from
Chicago, one of those experts, CFO Laurel Giarrusso, told him that if
he did indeed want to stay at the helm of his budding organization,
Mark would need to once again place his trust in someone else—in
this instance delegating someone to take on the responsibility of making these weekly sales calls so that he could seek additional investment
capital—and the budding entrepreneur heeded her advice.

"She was right, of course," Mark said of the conversation. "I was
going out on the road every week, working incredible hours. And it
was good because we were making deals, and I learned a lot about the
deficiencies of the product, hearing what else should be added based
on what prospects and customers wanted out of it. But we had all but
used up that first big round of financing to get to that point."

Mark would need to stop selling clocks and start selling Kronos . . .
again.

But whom should Mark turn to? Who could Mark entrust with a
function for which he was ideally suited—one in which he had absolutely excelled to this point?

Luckily, the answer to that question was easy. And shortly thereafter, Mark was briefing brother Aron, who was adept at installations,
making demos a snap. In addition, Aron had also already demonstrated true sales ingenuity with his innovative way of creating leads
from the Yellow Pages. Now Mark needed to entrust his youngest
brother with the front end of the sales process so the elder Ain brother
could recapitalize the company.

Aron would travel to the Carolinas, and be successful. He'd journey to Houston, and also hit pay dirt. Along the way, Aron, who had
worked hard since his arrival at Kronos to find ways to become increasingly valuable within the company, stepped into a function whereby he
could more directly affect Kronos' bottom line . . . and his own.

"I hadn't done much traveling for Kronos at that time. I went to
New York a lot for installations and maintenance. And of course, I was
doing service calls around Massachusetts. But I didn't count that as
traveling," recalled Aron. "But I was young and I knew the product, so
this was exciting."

Aron ran into the same pushback that Mark had encountered . . .
the issue of price. But like his older brother, he was able to stress the
value proposition the Kronos clock brought to the equation. And subsequently his record of success was becoming equal to Mark's.

These first couple of trips . . . these first few steps in a journey of a thousand miles, would beget others for Aron, as he would rise through the ranks to eventually open the first Kronos satellite sales office, head up sales, lead the service organization, serve as chief operating officer, and ultimately succeed Mark as CEO. But at the time, he was simply looking for a way to step forward when an opportunity was presented.

Subsequently, Mark's quest for additional funding would likewise prove fruitful for him, for Kronos, and for those wise enough to invest in the company. It took some time—over three months. But not only was Kronos recapitalized, but it was once again done without Mark needing to relinquish control of the company.

All it took was long hours, hard work, dedication, and the fact that Kronos had indeed created a truly better mousetrap.

Chapter Nine

THE GIANT PAID THEM NO HEED . . .

The value of an idea lies in the using of it.

—Thomas Edison

During Mark Ain's search for the right product upon which to found his company, he sought an industry that had not yet reaped the benefits of incorporating the emerging technologies of the day. But though that was his principle criteria, it was perhaps a secondary ingredient to his game plan that ultimately allowed a fledgling company like Kronos to survive long enough to thrive.

The realm into which he would seek to make inroads should, he believed, be dominated by a behemoth: an engineering and manufacturing titan so enormous within its space that it would pay little heed to a tiny niche player attempting to reinvent how things were done.

The realm of timekeeping in the United States turned out to be just such a domain. Clocks were manual, and so was the laborious process of tabulating the time represented on time cards. At the top of the heap sat a giant—one blissfully unaware of Mark's aspirations to play David to its Goliath.

Ironically, the behemoth in question, the Simplex Time Recorder Company, had been born of a technology-based innovation of its own. Edward A. Watkins, who was in charge of the engineering department of a then-prominent furniture maker in nearby Gardner, Massachusetts, had invented and patented one of the first practical time clocks back in 1894. Watkins' device was considered one of the earliest information technology devices ever deployed in a workplace. Almost a century later, and after a 1950 acquisition of a time clock division within IBM, Simplex was the gold standard in timekeeping in the U.S. And a third generation of the Watkins family was at the helm.

Mark didn't stand in front of Simplex as he loaded his sling with a stone. But he might as well have.

"Quite a few of the salespeople we hired had been selling for Simplex," he recalled. "In fact, when we went to a new city, we'd purposely find Simplex salespeople and convince them that ours was the clock they should be selling."

The process worked. But even with that, Kronos was barely making a dent in either the industry or Simplex's sales force. The giant slept on, well fed and snoring. Or did it?

Fresh from securing the funding needed to stay in expansion mode, Mark had decided he would open up investment opportunities for family and friends as well as the handful of employees who made up the Kronos workforce, among them Larry Krakauer.

By this time, Krakauer had dedicated extensive time toward refining the Kronos offering. He'd played lead in the debugging of the earliest version of the time clock. And he'd spent more time working than he'd planned when he first decided to take the leap and join the team. But he believed in the game plan Mark had established, so much so that he'd discussed with his wife the possibility of taking Mark up on the offer to invest in Kronos, a quantum leap of faith for someone with a young family.

Krakauer attended a due diligence meeting, a final gathering of potential investors, and was happy to be ponying up into a venture he felt held potential. That is, he was pleased up to the point when another investor-to-be arrived with a flyer espousing the virtues of a creation from time clock Goliath Simplex.

"As the meeting started," Krakauer would later recall, "one of the early investors arrived with a flyer advertising a new electronic time clock, called the Step99, from Simplex."

To that point, the prevailing notion within Kronos was that Simplex was a purely mechanically oriented company, one that was blissfully unaware of the power of microprocessors or the seismic shift about to be promulgated by modern computer technology.

"I looked over the flyer," Krakauer recounted, "and immediately saw what appeared to be a competently designed electronic time clock."

Turned out that on top of its U.S.-based facilities, Simplex also had a development facility in Germany. There, Simplex engineers were well versed in microprocessor-based design. And, having produced and successfully sold the Step99 (or Schritt99 as it was called in Germany) in Europe, Simplex executives, still mainly unaware of Kronos, intended to market the solution in the United States.

"As I sat in the meeting, I thought to myself, 'We're dead!'" laughed Krakauer. "Simplex, after all, was the 500-pound gorilla of the time clock business. Most of the time clocks in the United States had stickers on them saying 'If you need to replace this time clock, call Simplex' followed by their phone number. They had thousands of salespeople across the United States. If they had a well-designed product, it seemed to me, we had absolutely no chance."

As Larry would put it, he suddenly found himself in a "ticklish" situation. He'd been offered the chance to buy Kronos stock. And, after some pondering on the home front, he'd accepted. The due diligence meeting was supposed to be a formality. But at that point, staring at Simplex's Step99 brochure, he was a man in conflict.

"If this information had been available *before* I was offered the opportunity to invest, I could have easily declined and nobody would've given it a second thought," he said. "But having already accepted the invitation to invest, I thought that to back out at this late point in the process might be interpreted as a lack of confidence in the company's prospects."

Of course, at that point, he *was* worried about the company's prospects! With the right product, and this Step99, at least on paper, certainly looked like that product, Simplex could eat up Kronos in much the same way the mythological Kronos, the Greek God of Time, had a propensity for devouring his children whole as snacks.

"If I pulled out at that juncture," Krakauer mused, "I wasn't sure I would have much of a future with Kronos."

Other investors in the room that day were likewise aware of the development. But there was a sense of calm in the room, with an overriding notion that competition was inevitable. It might have been better if said competition came in the form of another start-up operating in an unlikely facility that smelled of burning steel and stale carpeting to mirror where Kronos was at that juncture versus what appeared to be innovation emanating from the lair of a worldwide leviathan like Simplex. But again, and perhaps bolstered by Mark's trademark confidence, or maybe the fact that the Ain family was likewise investing, most, if not all, held firm.

"To my knowledge," Krakauer reported, "nobody backed out of their investment based on this late news . . . so I just swallowed hard, signed the documents, and handed over my check."

For his part, Mark remained steadfast in his belief that Kronos was on solid ground. The current iteration of the Kronos time clock, with Krakauer's tweaking, was ready to disrupt the industry. And his extensive, nay exhaustive, customer research told him that the immediate threat posed by Simplex and the Step99 was not nearly as imminent as others in that room that day might have thought.

True—rolled out in Europe, the Step99 was a well-designed unit that performed admirably in the field, and subsequently sold a more-than-decent 10,000 units. But as Simplex tested the waters in the U.S. marketplace, a cultural flaw was exposed. The unit had buttons!

When employees needed to clock in or out on the Step99, one of four specific buttons needed to be pushed to properly record the punch. These four buttons indicated "Entering for the Day," "Leaving for a Break," "Returning from a Break," or "Leaving for the Day."

Pick one of four buttons. Simple. But the Kronos time clock did not have buttons for one simple reason: Mark's many discussions with potential customers, as set up by sales channel partner Interboro, said American workers would just say no . . . or worse.

"What we learned," recalled Mark, "was that American workers were both distrustful of all things electronic at the time and particularly skeptical about the clocks."

But the clocks were a success in Germany?

"Apparently, when you tell German employees what they need to do, they say '*Jawohl*' and do as they are told," reported Krakauer, in reference to a Germanic military term that indicates that one has heard and will immediately comply with an order.

"When you tell Americans which buttons to push," he continued, "they promptly push the wrong button, sometimes deliberately, to see what kind of chaos they can cause."

Needless to say, with an objective of being accurate and a time saver, chaos was not an added feature listed on the pamphlet seen at the due diligence meeting that day.

So why didn't Simplex pivot, instruct that same German engineering team to create a version of the Step99 (perhaps to be called the Step100?) that mirrored the specs of the Kronos clock, and simultaneously wipe out both the upstart company and Larry Krakauer's life savings?

"Why *didn't* Simplex wipe us out?" Krakauer asked rhetorically. "The fact is, they should have. All they needed to do was to have their German subsidiary make a slight change in the design of the product for the American market. But they didn't do it."

Two theories would become folklore at Kronos. Likely both were true to a certain degree. And each came back to Mark's initial premise: the giant—big, fat, and well fed—felt itself to be too big to fail.

One school of thought held that Simplex felt that microprocessor technology was too expensive. Yes, they could see these new clocks, but the price point seemed high for the industry. Of course, this was true.

"Our early clocks cost about $5,000 each. And they were compet-ing against mechanical time clocks that sold for less than $500," Mark said.

"There was definitely pushback on the cost," brother Aron Ain would say about his early sales calls.

"It's possible that the Simplex management's lack of familiarity with microprocessor technology did also play a role," said Krakauer. "The electronically savvy Kronos management had confidence that the price of microprocessors would decline sharply over the next few years. If the Simplex management was less informed in such areas, they might have assumed that electronic time clocks would remain. over-priced, compared to their purely mechanical brothers, as they were in 1979."

"The technology was definitely a stumbling block," said Mark. "But the truth is that their salespeople just didn't want to sell electronic time clocks. They were comfortable and doing very well with the mechani-cal version. So, after they went out and tried once or twice to upsell their clients, they went right back to pushing the mechanical version."

"Salesmen like to sell what they know," said Krakauer. "As many in business have painfully learned, your greatest strength can sometimes be your greatest liability. The Simplex salesforce was highly proficient at selling *mechanical* time clocks. They were not knowledgeable about electronics, and this microprocessor-based, German-designed unit scared them."

The Simplex sales force, thousands strong and a major asset, thus became an unrecognized Achilles heel. Simplex, convinced the United States would never become interested in electronic time clocks, not only withdrew the Step99 from the American market, it later sold the German subsidiary that had created it.

Simplex—the giant—had been in possession of the solution. But had opted to not utilize it.

Over the next few years, as the cost of microprocessor technology dropped, and Krakauer and the rest of the engineering team enhanced the ease-of-use, range of features, and accuracy of the Kronos clocks, tiny Kronos began to first creep up on and then roll over the giant..

Twenty-four years later, in 2003, Kronos would finish the play by buying what little was left of the Simplex time clock empire. By this time, and with the giant vanquished, Kronos had become a worldwide force in the collection of time. And Larry Krakauer's early investment would have him well positioned financially to begin pondering a more-than-comfortable retirement.

Chapter Ten

SOLVING A BIG PROBLEM . . .

Size matters not. Look at me. Judge me by size, do you?

—Yoda, Jedi master

Giraffes are the world's largest terrestrial animal. Adult males can stand 18 feet tall and weigh over 3,500 pounds. Their prominent necks alone can hit 600 pounds on the scale. And they are capable of killing an adult male lion with a single blow from any of its four legs.

Apex predators like the lion, and sometimes even a crocodile, will sometimes kill a giraffe, but they generally avoid healthy specimens unless absolutely necessary for fear of fatal reprisals.

Enter the lowly tick. Less than the size of a nickel, ticks attach to giraffes and burrow beneath their thick coat and begin a process of slowly sapping the animal's strength and eventually causing an infection that in and of itself doesn't necessarily prove fatal. What proves fatal is being slowed and weakened, something that does not go unnoticed by the top of the food chain.

As noted previously, Mark had consciously sought out an industry giant that had not yet figured out how to work the emerging technology of the day into its strategy. And so large was Simplex that it barely, if at all, noticed as Kronos burrowed into the time clock business and began slowly sucking customers away.

That giant, blissfully unaware of its imminent doom, continued grazing on the leaves of the industry!

In business, giants come in varying shapes and sizes. The apex predators are easiest to spot. But it's the underlying issues—giant problems, if you will—that often determine a fledgling company's trajectory toward success or failure, as measured, ironically or not, by metrics known as upticks and downticks.

Mark didn't raise a victory flag as Kronos lifted buoyantly away from the launch pad. Rather, with product shipping and accounts receivable, and with financing under control, he set his mind toward improving every aspect of the operation from procedure through product. It was time to take a good look inside and see if there were any issues that needed preemptive action before turning bigger.

"We'd done very well bringing the right people on board to do what needed to be done to get us started," he said. "But not every hire was a perfect fit. Sometimes, someone did not work out."

Of note at that point was the current head of sales. Mark felt that this individual wasn't hitting the level of effort he demanded from everyone. And upon further evaluation, he felt it was time for a change.

"Just wasn't the right fit," Mark reiterated.

There was a slight hitch. The gentleman in question was a former Division 1 football lineman. In other words, he was a giant in his own right.

A fellow named Samuel Clemens once wrote a very memorable and oft-cited axiom. "It's not the size of the dog in the fight. It's the size of the fight in the dog."

Words that ring with a certain sense of bravado, a notion that physical size doesn't always matter. But they were written by Clemens under a pseudonym! Clemens didn't puff his literal chest out and boldly make this claim. His alter ego Mark Twain delivered the message between puffs on his tobacco pipe.

Yet, where others would have sent an emissary with bad news, Mark stood as tall as possible and took the meeting solo.

"Was I a little worried?" he pondered. "Truthfully, yes . . . at least a little bit. But I wouldn't send someone else to do what I knew had to be done. And more than that, I wanted to show anyone who'd worked for me, whether they worked out or not, the respect they deserved."

In walked Mark. And this giant was gone. But far from feeling a sense of victory, now there was a void. Mark had big, critical, double-wide shoes to fill.

But Mark didn't need another pair of size 15 shoes. What he needed was someone who carried himself gracefully, yet forcefully. He needed someone who understood the value of technology, which was still a younger person's game . . . on a peer level with key decision makers within prospects, who tended to be older than most of the employees at upstart Kronos.

Someone sharp. Funny . . . but in a higher brow sort of way than you'd find in a locker room. Someone who could be intimidating intellectually rather than physically. Someone who could think outside the box, because the Kronos sales model, as configured at the time, was revolutionary on several fronts.

He needed . . . Pat Decker.

Decker had come from the other side of the Charles River from MIT, from Boston University. Immediately, he'd been hired into a sales position by the Boston office of the ever-growing computer colossus IBM, where he toiled for five years. He'd then joined a company called Centronix Data Computer Corp., a pioneer in dot-matrix printing, and managed their worldwide dealer network before a short stint with Exxon was followed by time spent working for Data General, one of the world's first mini-computer firms.

Affable, intelligent, and experienced in the worlds of both emerging technology and dealer networks, Pat had one other intangible going for him—he had gone prematurely gray!

"I did," he would later confirm with a chuckle. "So, I looked a bit older than I was."

"Pat," said Mark, "checked every box. He had the experience, and the temperament. And he did look a little older!"

A colleague had pointed Pat to Mark and Kronos. And, given the timing, things looked right for both parties.

"I told him we had somebody in this role before and he didn't work out," detailed Mark. "I described Kronos. And told him what we did and didn't know."

"I'd done quite a few things in electronics, and in distribution and sales, and dealer networks," remembered Pat. "But it was a time that technology was just starting to blossom in the Route 128 beltway and I'd come to the conclusion that I wanted to find a business that had a very targeted product aimed toward a defined marketplace where we could make a difference. And Kronos really fit that bill.

"They had the technology working in an area . . . time clocks . . . which nobody was really paying much attention to," said Pat. "Me? I thought managing labor would ultimately be a very big business. That's what attracted me to Kronos. So, from then on we really sort of grew up together in the whole area of developing the time clock business and then scheduling and management."

Less than three months earlier Mark had sent brother Aron to Chicago to start the company's first sales office. Shortly thereafter Kronos had similarly opened a Los Angeles–based sales office. Both operations were gaining footholds while time clocks were being sold through dealers in much of the rest of the country.

"So, we were doing both," said Pat. "We were managing the dealers and distributors and we were learning how to sell the products ourselves through those direct offices in Chicago and LA."

The plan was to properly organize and incentivize the dealers, while continuing to learn the best way to market a $5,000 time clock in a $500 time clock industry. Big challenges and the potential for a high reward for the effort lay ahead. But first Pat showed Mark and Aron some sales savvy.

"The first really big customer we struck a deal with was Marriott," Aron recalled. "And they had pretty much decided to go with us. But there was still the question of price. We met at one of the hotels. And they asked about price and Pat told them $5,000. And the Marriott folks were saying 'Wow! $5,000 for a time clock is a lot of money.'" So, we go back and forth over price and finally someone suggests we break for lunch. And someone brought out the menus to order room service. Pat opens up the menu and looks at it and says, "Eight dollars for a hamburger . . . that's a lot of money!' and negotiations were done."

"That was a big deal," recalled Pat, "because they were national. That made us national."

Coming on the heels of a deal with a Wisconsin retail group with 30-something stores, Kronos had more momentum than ever before. And the team set about setting a course for more milestones. And, like the product, that meant continuing to innovate and evolve the Kronos sales process.

Pat saw what Mark had established to this point.

"What was happening in the general marketplace," Mark recalled, "was that anybody selling products would sign up three or four dealers in an area and hope somebody made some sales.

"I didn't think that was a good strategy," Mark said. "So, with this marketplace, we decided that we would basically give dealers a protected territory . . . give them sales quotas to meet . . . and then try to work with them to help them succeed."

Because Kronos had its own sales offices in selected areas like Chicago, the leadership team was able to prove which sales strategies would work best. They would then pass along those best practices. That strategy worked for several years and helped the company get launched.

But, as Pat detailed, quite a few dealers came with a finite shelf life, given the vertical trajectory of Kronos sales.

"As we found out over time," Pat said, "dealers would only invest so much money and would only change their business model to varying degrees to be able to be successful with Kronos. At the same time, we had two branch offices, and especially Chicago where Aron was doing a great job with a great team, that would show us what the potential was for selling products.

"So, we'd get lists of companies with more than 100 employees in every county across the country," Pat continued, "and we'd set our quotas for the dealers based on what we ourselves were experiencing within the branch offices.

"Some of the dealers were very successful," he added. "They were very willing to change their business model and follow our lead about who to hire and how to manage them. Some of them didn't. The ones that did became longer-term partners. The ones that didn't would at some point be told that they'd been missing quota so we are either going to find a new dealer, or we'll start-up another branch office."

"We wanted sales partners who would embrace our way of thinking," said Mark. "We gave dealers big discounts when they exceeded quota, so the more they sold the bigger the wholesale discount would be. But just as we at Kronos were taking a good share of our profits and rolling them in R&D, we expected the dealers to reinvest some of their profits into their operations. They could use that additional profit to expand their businesses by hiring better-quality people, people who were more suited to contributing to success."

"The whole distribution model morphed to a point where, for several years, we had no direct sales offices in the majority of the important territories in the United States," Pat recounted. "But that really helped the company take off. At the time the idea really made sense because those who produced with better-quality people were more suited to succeed. And again, as Mark had foreseen, the combination of the direct sales lessons we were learning that we could transfer to those people of partners who were willing to invest in their channels properly and work at it translated into success for both parties.

"A lot of those dealers," Pat added, "became very successful over a ten- to twelve-year period."

Things were taking shape, yet changing at the same time, something that was absolutely necessary given the embryonic stage Kronos was in. The company was now known nationally, particularly by those in the "back rooms" of organizations where employee time was kept and accuracy was both an issue and an advantage.

"It was a fun time, seeing the business grow and watching so many people succeed," said Pat. "And it was very collaborative. Mark and Aron would have ideas on what we should do. I'd have ideas and they were always open to those ideas. I always felt that we had a real mission.

We knew that we were doing something unique. So, we took Kronos forward together."

Over time the whole sales model would change. Kronos would slowly acquire dealers and their territories and bring those territories in-house as the internal Kronos sales staff began to swell to a point whereby almost all sales were handled by what would become a global sales force. That change was imminent. But to Pat, who would eventually rise to serve as the company's president and be part of a management team that continuously drove Kronos to new frontiers and greater heights, best practices and a culture built on attracting the right people were constants.

"We hired a lot of good people," he said, "and a lot of those people stayed for a long time. Some are still with the company literally decades later. So, we all grew together. And there was always this really strong work ethic in the cult of the company. Everybody was dedicated. Long hours? It was how we got the job done. It was never about how long anything took. And the people that tuned into that and really appreciated that approach and wanted to be a part of it found a way to succeed."

Spoken by someone who didn't physically appear to be a giant. But Pat Decker, a giant of a man nonetheless, more than filled the shoes of his predecessor, and then some, at a time when Mark and Kronos needed someone uniquely like him to step into the picture.

Figure 1. Mark enjoyed his visits home to spend time with family. Here, he poses with youngest brother Aron over dinner. Decades later, Mark would step down as CEO of Kronos, leaving the role to Aron.

Figure 2. Kronos employees outside the company's first office in Brighton, Massachusetts, gathered by the station wagon that would take the first shipment of electronic time clocks to New York City. Mark Ain is at far right.

Figure 3. One of the first batch of Kronos' electronic time clocks, which sold for over $5,000 each. Kronos sales competed against the $500 manual clocks manufactured and distributed by Simplex and other companies. But the efficiencies and cost savings that Kronos provided offset the higher cost.

Figure 4. This package of clocks, a personal computer, and floppy disks holding Kronos' proprietary software represented a first generation of timekeeping solutions capable of communicating time data directly through a PC.

Figure 5. An early Kronos promotional parody featured Father Time realizing that even his "time" had come with the development and emergence of the company's first electronic time clock. The patented time cards pictured were key to providing the fledgling company with a forecastable revenue stream.

Figure 6. An early 1980s picture in Kronos' lone boardroom in Brighton as CEO Mark Ain prepares for an early investor meeting. Mark's approach to raising operating capital for the start-up version of Kronos ensured that the company was positioned to reinvest early revenue in constantly improving its product line.

Figure 7. Though ties were seldom part of the dress code at Kronos, this early 1980s formal photograph features the company's engineering team. At far right is Larry Krakauer.

Figure 8. Early 1980s photograph of the sales and support staff. Alice Ain is far left. Pat Decker is seated at center. Mark Ain is to the right of Pat.

Figure 9. By the mid 1980s, Kronos and its timekeeping solution had earned a seat at the executive table.

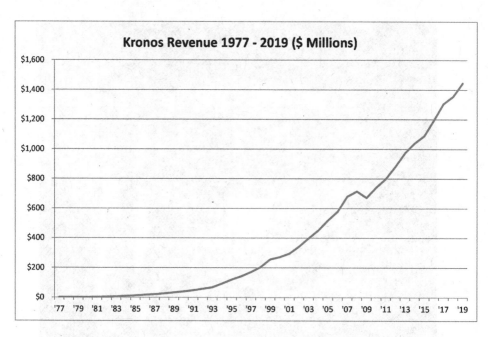

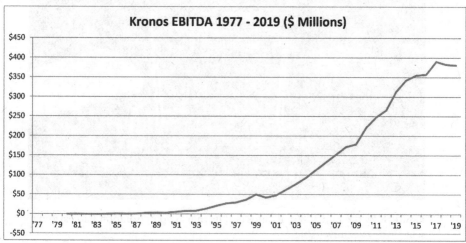

Figure 10. Kronos' records for revenue and profitability growth ascended for decades.

Figure 11. Mark Ain in an early 1980s public relations shot. As Kronos solutions began to show value, industry magazines started writing stories about the fledgling company with a new, high-tech solution to a persistent business challenge.

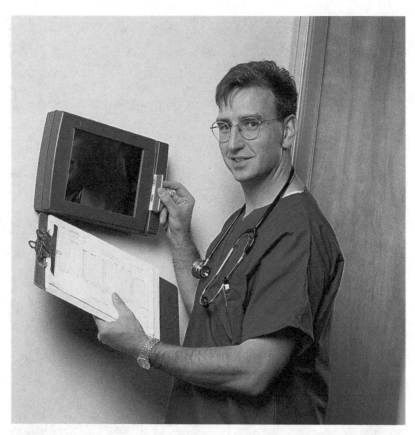

Figure 12. During the economic crisis of the mid-1980s, Kronos's next-generation time clocks took hold in nursing homes and hospitals; keeping the company afloat as many other organizations floundered.

Figure 13. By the late 1980s, Mark had honed his investor presentation skills enough to add funny captions to the shots documenting his efforts.

Figure 14. When Kronos was invited to ring the opening bell for the NASDAQ in 2002, he was happy to be able to share the moment with his mother Pearl.

Figure 15. Carolyn and Mark, along with Pearl and Jack Ain, just before ringing the opening bell for the NASDAQ.

Figure 16. Carolyn and Mark Ain projected over Times Square in New York City as part of the NASDAQ celebration.

Figure 17. Mark had to ring the bell with his left hand because he had just had surgery on his right shoulder.

Figure 18. Aron Ain, Mark, and Paul Lacy at the opening of the company's new World Headquarters in Chelmsford, MA.

Figure 19. The five Ain siblings gathered for a group shot at a family event. From left, Aron, Brent, Alice, Mark, and Ross.

Figure 20. A group photo of Mark and Carolyn with their immediate family at the celebration of their respective 75th birthdays.

Chapter Eleven

ANOTHER TALL ORDER

If there is one secret of success, it lies in the ability to get the other person's point of view and see things from that person's angle as well as from your own.

—Henry Ford

In the world of sports, the best teams are not necessarily the ones that contain the greatest number of star players. For every Tom Brady and Rob Gronkowski, for every Larry Bird and Kevin McHale, there were many players whose blood, sweat, and tears made for championship efforts but who never became household names to those who were less than hardcore fans.

As a company, Kronos played that unheralded, backroom role for most of its history within the sport of business success. Customers who utilized the Kronos time clocks, and later the myriad of solutions that contributed to their running better, more cost-effective businesses with a healthy and rapid return on investment, knew the name. But the average consumer, upon hearing the name Kronos usually replied "Who?" unless they were Greek mythology buffs.

Likewise, within Kronos itself, there were marquee names, but there would have been no championship run without the guys who did the blocking and tackling or the boxing out and rebounding.

Mark understood this as clearly as he knew right from the start that he was not going to be the person who physically designed those initial time clocks. Mark had the idea and the business savvy. But without a Larry Baxter in the mix, the idea would have stayed where most earth-altering, life-changing ideas usually reside: in someone's head.

Now, with product shipping, a sales network comprised of both dealers and in-house staff selling and increasing revenue, and a burgeoning stable of engineers charged with perfecting and reimagining the clocks, Mark took on the tall task of finding someone to ensure that Kronos' own accounting was taken care of properly.

The search had proved fruitless to the point where the ever-frugal Mark had opted to engage a search firm to assist in the quest. But that too, had not borne fruit. That is, until the day he looked up, both figuratively and literally, and laid eyes on a six-foot, seven-inch, former Farleigh Dickinson University basketball player named Glenn Bolduc.

Ironically, it was a meeting that wasn't supposed to happen. As with other critical hires in those early days, Mark's ideal candidate would need to be a hybrid. More than a number cruncher, he sought a senior accountant type with experience writing business plans . . . someone who would be good with the books of the fledgling company, but who could also write and fiscally help defend business proposals.

More than that, a candidate who could pass Mark's standards would need to also be willing to join a start-up with, at best, meager-looking offices and an environment where each employee, regardless of level of expertise or experience, was also willing to man a broom or physically move a time clock or two during the course of an overly long business day.

Glenn Bolduc, at twenty-nine years old, was nowhere near senior level. He had a strong cost-accounting background, having started his post collegiate career with accounting giant Price Waterhouse (now PriceWaterhouseCoopers) before moving into the private sector as a controller. The Greater New York native was actively looking however, as his current company had been faltering to a point that the head of the company, a man Mark knew, had actually given Glenn permission to look elsewhere.

That didn't put Kronos on the horizon just yet, though.

The former college basketball forward was far younger and even less experienced than Mark envisioned, but a chance decision by the headhunter trying to fill the position eventually made Buldoc stand above the rest.

"The headhunter, whom I had worked with in the past, contacted me and asked me to do him a favor," Glenn recalled. "He had the exclusive on Mark's CFO position. He said 'You're not a candidate' for the job. You're too young . . . not enough gray hairs. But I'm having difficulty selling it.'"

The favor was that the headhunter would send Glenn to Kronos so Glenn could report back with a clearer picture for what an ideal candidate would look like. He was told "the guy that is filling the job (Mark) is a smart guy, but everybody I send up there just doesn't seem to match up."

His initial interview wasn't with Mark however. It was with Deane Farnsworth, the VP of manufacturing at the time.

"I could tell right away he wasn't expecting a twenty-nine-year-old to be walking through the door," Glenn chuckled. But the interview actually went pretty well and the next thing I know I'm getting a phone call and I'm being told Mark wants to meet me."

Whereas the position had been a tough one to sell from a head-hunter's perspective . . . those with the right level of experience either didn't want a start-up in general, didn't see the upside of the value proposition of a time clock that cost ten times more than a traditional one, or their personalities didn't mesh with Mark.

Though not as seasoned, Glenn immediately understood the value Kronos could bring as a strategic partner with its customers.

"I was a controller at the time," he recounted. "And I handled pay-roll. And the calculation of hours worked was both a pain in the neck and subject to error all the time.

"And," Glenn continued, "if the error was in favor of the employee, you never heard about it. But, if the error was against the employee, you could be facing a lawsuit. So, it was important stuff."

Mark had gotten a favorable report from Deane. More important, Mark knew Glenn's boss as a person of high standards who would not advocate for anyone who didn't earn that right.

"I never forget Mark's words," Glenn said. "Mark said, 'I see you work for Don Cramer. I know Don and he is a tough taskmaster. And if you can live in his world then you've got to be a decent worker.'"

"It was, again, an instinct call," Mark remembered. "Glenn wasn't necessarily who I thought I was looking for. But in the end, he was the right person for the job."

In large part, that decision came down to the fact that Glenn walked tall through the door (perhaps stooping a tad at the entryway), and rather than pander to Mark, he gave him concrete feedback.

"Mark didn't know the balance in the Kronos checkbook to the nearest ten thousand dollars!" said Glenn. "And I said 'Mark, as CEO, you should get a cash report every single day. When you go to bed at night, you need to know how much money you have in the bank!'"

Mark liked the tall kid's attitude. And he saw immediately that what Glenn lacked in experience, he made up in savvy and potential. But before he would push the plunger, he had to make one important phone call.

"In the next interview Mark went so far as to call my wife to ask her how she felt about me taking this position," said Glenn. "He told her that it was going to mean long nights . . . that this was a start-up."

Though initially taken aback, Glenn quickly realized that in Mark he had found someone who both had a roadmap for creating his own path to success within the ongoing technological revolution, but also cared that the people he brought on board would be happy enough to put in the kind of effort he felt was necessary to achieve success.

"You know how some people will describe someone as 'he got game'?" Glenn asked. "Well, Mark is Vince Lombardi, Bill Belichick, and Bill Parcells all rolled up into one. What I recognized with Mark is he had this ability to think everything through to the future. Mark had that ability to manage a dynasty the way Parcells did with a couple of teams, the way Lombardi did with the Packers and Belichick did with the Patriots.

"So, at first, I got mad about him calling my wife," said Glenn. "In fact, I think I told him to never call my wife again without asking me first. But after I went to work for him, I realized that I was working with somebody who was one of the smartest people I would ever meet. And, to be honest, this was sometimes incredibly frustrating, because he was typically at least five steps ahead of everybody else. And he's also one of the most impatient men in the world. But I came to realize that the call was Mark being Mark. He was Lombardi or Belichick or Parcells making sure that the dynasty that he was creating would survive . . . and that he had the right player at the right time."

So, Glenn came on board at Kronos. And Mark immediately started getting his daily cash reports. And once a week the two would meet to discuss what cash was needed and when to continue to grow the company.

"I think we did fourteen finance rounds in my first six years," recalled Glenn, "which sounds like a lot, but we raised money when we needed it and we always raised money at an advanced valuation. We never had a down round . . . and that's pretty remarkable if you think about it!"

Glenn credits Mark for holding the line on the Kronos valuation and on being astute in the art of the deal as illustrated by his early trip to Chicago with Mark to close the previously cited big-at-the-time deal with Sears.

"Mark would bring me on these trips because I was six-foot-seven, and size can be intimidating at times," he laughed. "But I remember this trip for a couple of reasons. First, I think we had to hold Mark down so we could put a tie on him. The man does not like getting dressed up. And, of course, this was Sears so they had a lot of stores where they needed time and attendance products so this was a big deal.

"Mark," Glenn continued, "knew our sales guy had promised them a big volume discount. But he was pretty quiet as they marched in . . . I think it was fourteen or fifteen vice presidents. And in a very soft but forceful voice Mark tells them that we're not going to be able to do this deal at the discount rate they're looking for. In fact, he told them, we really need to get a higher rate."

Mark's message took the salesperson and others by surprise, but not Glenn.

"I had the benefit of being with him in the main office every day and hearing what his strategy was going to be," he said. "That's the thing about Mark. He would always bounce his strategy off people. And it would run the gamut. He'd go 180 degrees. It would be the most ridiculous stuff to the most commonsense stuff. And you knew that the answer was somewhere in the middle. But Mark always wanted to get all of the alternatives on the table so that we were always in the best position to make the right decision."

Knowing what was coming was one thing. But seeing it transpire in real time left a lasting impression on the young CFO.

"What an amazing lesson I learned right there in about fifteen seconds!" he said.

It wasn't a complete victory though, at least not right away.

"We installed about forty clocks with them," Glenn said. "And then we discovered a software bug and had to go back and upgrade the memory in every one of those clocks. But after that, Sears became a great reference for us."

With cost accounting and internal consulting as under control as it could be, given Kronos' start-up status, Glenn turned his attention to honing his business-plan writing skills as part of the function of securing investments in the company.

"Keep in mind," he said, "that I was still just twenty-nine, thirty years old when I was doing this. And I did not have the benefit of going to business school. I came right out of college and I went right to work for Price Waterhouse. That was my business school.

"I wrote one business plan, and it was, to be kind, terrible," he recalled. Luckily, another member of the board saw the business plan and came to his aid.

"He spent four or five hours with me," he said. "And bear in mind we didn't have word processors back then. So, the revised document was a tape document. And by that, I mean we had scissors and Scotch tape out, cutting and pasting a plan together."

Rudimentary to be sure. But the resultant effort ended up bearing fruit. It was sent to a London-based venture capital firm called Churchill International, then one of the largest venture capital firms in the world.

"We actually got to meet the CEO of the firm and he paid us the highest compliments on that business plan," Glenn recounted. "He came in and he said to us 'It's about time I get a document that doesn't have all the fluff.' This was a nice ten- to twelve-page document that went right to the point about the product, the market it served, and the vision Kronos had for that market.

"In the end," he recalled with some degree of pride, "we got a $750,000 investment from Churchill International."

Glenn would quickly point out that he was proud, but realistic.

"By the way," he added, "Mark could have written a business plan at any time if he wanted. I know that now, whereas I didn't really know it then. But there was no pride of authorship with us. And he was willing to have me get better at it by doing it."

Which didn't mean that Mark wanted Glenn to have a hand in every project at Kronos.

"The nickname that Mark hung on me was Dr. J, because I wanted to be involved in everything," Glenn reported. "But at one point in time Mark sat me down and said 'I need you to do the things that you are good at.' That was a hard lesson for me to learn, but I eventually learned it. You let the people who have expertise do what they're good at because doing so means you end up with a good product. And that's true whether you're writing a business plan or building a time clock."

Decades later, as Glenn looks back, he relishes his primarily behind-the-scenes function, making sure that Kronos finances were buttoned down and his business plans were accurate and concise. He's grateful for Mark's attitude with regard to hiring . . . that sometimes the best fit for a round hole might be a square peg after all. He's proud to have been a role player in a high-tech success story despite his initial lack of high-tech background. And he's forever glad that he was sent to Kronos by a headhunter who was certain he wouldn't get the CFO job, but sent him anyway.

"Honestly," he said. "I didn't completely understand the whole high-tech craze. I don't know that anybody did back then. But in my time at Kronos, I was able to see things come together.

"More than that," he continued. "There was a personal element to all of this, and that was because of the way that Mark managed . . . the way he sought input . . . the way he was always fair. He put some good people in very good positions not just at work at Kronos, but in their lives, myself included.

"At the end of the day, Mark was always very real," Glenn said in summary. "I don't know that that's the first impression most people got. But after you got to know him, you understood that because, at the end of the day, he always owns up to whatever has to be done. Mark was as tough a taskmaster as I ever worked for, but he was also very fair."

And as for his role . . .

"I'll leave you with one thing," said Glenn. One of the greatest baseball players ever . . . Mickey Mantle . . . do you know what's written on his monument? It reads, 'A Great Teammate.' I wouldn't mind at all if someone would refer to me as a great teammate. That would be the greatest honor I would ever get. I think that's all I would ever want to be . . . a great teammate."

Chapter Twelve

ESPOUSING THE VIRTUES

There is only one thing worse than being talked about . . . and
that is not being talked about.

—Oscar Wilde

As the 1980s kicked into gear, the financial mood was upbeat as the economy was slowly turning around, essentially signaling what would come to be known as a decade of greed.

Madonna and something called rap were making waves in the music industry, while music videos were coercing people to "watch" their music as everyone wanted their MTV. And the first generation of personal computers, known simply as PCs, began appearing in residential homes.

In Boston, high technology was being advanced around the clock. But despite the presence of a bona fide revolution in the way workforces were being managed, what wasn't part of the everyday vernacular was the name Kronos.

True, Kronos was reinventing an industry. But the sales hill was a steep climb, not because the company's emerging technology wasn't a game changer, but rather because the message was not necessarily formulated to engage and enthrall.

Enter yet another all-but-unlikely key Kronite, as employees would come to be called, named Mary Jane Conary.

"Mary Jane really put the company on the map in terms of positioning," Mark Ain would later divulge. "What she did was define what Kronos could do in corporate terms."

Kronos didn't necessarily need the brand-name recognition of Sony or the buzz that surrounded MTV's music television. Mark needn't displace Madonna or Mr. T as cultural icons. But the word did need to spread that the new time clock's harnessing of the emerging technology of the day was opening managerial windows into workforce management in a way that the 100-year-old technology that Kronos was both displacing and replacing could not.

Kronos had held its own during the economic slowdown, depending heavily on building an early reputation with a few key retail and hospitality arenas while focusing on gaining traction within a relatively recession-proof industry: nursing homes (because, no matter the financial climate, people still must age!). The nursing home experience had pushed the Kronos engineering team to further improve the product. And with the advent of the PC, it was time to embrace and popularize the concept that what Kronos was doing was about much more than placing a microchip in a time clock.

Spreading that gospel fell to Mary Jane, a former newspaper writer turned ad agency rep. Again, finding the right person in the right place at the right time, Mark had interacted with Mary Jane when she had Kronos as a client of her ad agency employer. He felt almost immediately that she would be integral to Kronos' future success.

"She grasped the concept immediately," said Mark. "It wasn't simply about keeping track of time worked. It was about using that data to manage labor, manage scheduling, and payroll, and HR."

Conary had put together a media tour for Mark that would take them to New York City to meet with a number of reporters from various trade journals as well as the labor editor from *Business Week*. True to her guidance, she consciously took Mark to meet with them versus the other way around.

"At the time, Kronos was not in an environment that was built to impress," Mary Jane recalled, referring to the original offices in the ironworks building. "They were in an environment that was set up to get the work done. With the ripped carpet and the dirty walls, it was one step removed from working in a garage.

"I wasn't impressed with the environment. But I was impressed with Mark," she continued. "He was so driven and focused. And I was also impressed with the team he'd put together. Pat Decker was on board by that point. And Glenn Bolduc was there in Finance. He seemed to have all the players in place by the time I got there. So, it just felt right to me. Everyone understood what the vision was and everyone was just working really, really hard. I observed that entrepreneurial spirit in spades!"

"We had a very successful trip," recalled Mark. "And I remember that the labor editor from *Business Week* asked us if we could give him some names of companies that had put our clocks to work. And he wanted companies that had union workers so he could speak to both labor representatives and company representatives to find out what both sides thought of them."

"We met with Bob Arnold from Business Week," confirmed Conary. "I had called and my pitch to him was that this was a company that was going to change the way labor was managed in America. The old time clock hadn't changed in over 100 years, but this was something that he should be aware of because it's going to end up being revolutionary."

According to Mary Jane, the initial reaction was not too positive. Arnold wanted references.

"Arnold was quite skeptical," she recounted. "He said to me 'C'mon, that's just a pitch!' I said 'No! No, it isn't.' So, he agreed to see us."

During the visit, Arnold was given those references. He wanted companies that had labor unions, to see how both sides viewed the clocks. Almost to his dismay, employers and employees alike were pleased with more accurate timekeeping!

"He ended up writing a fabulous article and that put the company on the map," said Mark. "In fact, right after that I remember being in the office on a holiday . . . Washington's Birthday . . . and the phone rang and there was this guy calling from South Africa. He'd read the article and wanted to know if his company could have the exclusive rights to sell Kronos time clocks in South Africa!"

Somewhat ironically, the *Business Week* article represented international exposure. But it was the series of smaller trade publications that added up to an even larger share of Kronos business.

"We also generated a whole number of articles from various trade publications that covered industries where Kronos was trying to sell the clocks, like supermarket news, and other specific industries," said Conary. "On face value, those don't seem so important. But the sales force could go to sales meetings armed with those articles that told decision makers that this is something you should pay attention to. And that got them in the door."

Results mattered to Mark. And Mary Jane was getting results. So, he decided that he needed her as part of the Kronos team.

"I think we had just completed a second round of media interviews in New York," said Mark. "And when we got on the plane to head back, I said to her, 'You're not getting off this plane until you say yes to joining Kronos.'"

Which didn't mean that selling Mary Jane on joining the team was a slam dunk. But she was intrigued.

"They were looking for a director of corporate communications and marketing communications," said Mary Jane. "But I was perking along very happily in my job at the ad agency. But if you know Mark, you know that he's very instinctive. He observes something about people and then acts on it. And somehow Mark saw that I got the concept of what Kronos was about."

Thus, sometime between the plane landing safely on the tarmac at Logan Airport and it taxiing to the gate, Mary Jane became the first director of corporate communications and marketing communications in Kronos' brief history.

Problem was, the budget did not exactly line up with the length of the title.

"We just didn't have the money to go out and do spectacular advertising campaigns or huge direct mail programs, although that was part of the long-range goal," Mary Jane said. "But what we could do was rely upon public relations, because public relations was cheaper."

Public relations meant countless hours spent selling the Kronos story, recounting the modest beginnings while stressing the far-reaching impact that this company on the brink was capable of having, not just within its own industry, but rather within the vastly broader universe of labor management within all businesses. The big picture mattered a lot, but selling it was more grind than glitzy big splash—a chip and charge instead of an overhand smash.

Not that this was bad. As Mark had vowed in many a sales call early on, as much money as feasible was going toward making the time clocks better and better and better, adding capabilities and improving both functionality and scalability. And that commitment had made Kronos a game changer for a seemingly ever-growing number of organizations across all industries. Businesses were realizing dramatic cost savings based on their investment in Kronos, while also contributing to better employee satisfaction, another perhaps less obvious contributor to a better bottom line.

This deepening pool of referenceable customers came with success stories that resonated with prospects. Essentially, the story was about how this company or that organization had been more successful as a result of better managing its workforce, with Kronos a key factor in achieving that goal.

The bottom lines there were very marketing friendly, too. True, the time clocks were much more expensive than the ex–gold standard. But the cost savings realized by not only eliminating manual time tabulation but also through error reduction translated into returns on investment that paid for the more expensive clocks in surprisingly short order. Spreading those success stories was the ticket forward. Keeping current customers happy, always a mandate via Mark, made gathering those tales a relatively easy pull.

So, researching, writing, releasing, and tracking those customer stories became, according to Mary Jane, "the backbone of the whole marketing effort."

"It was just a matter of speaking with customers . . . getting their stories . . . and then selling those stories so they would be written in magazines," said Mary Jane. "So, we presented a huge number of stories across all industries and that got the early word out because management and employees were both seeing positive benefits from Kronos."

The approach also came with a multiplier effect. True, the Kronos story was a classic one: born from the entrepreneurial drive of an MIT grad in the midst of a groundswell of technological advances that were changing the world in a multitude of ways; created by a group of friends, first in basements and then in an ironworks plant; staffed by work-all-day-every-day types who were individually and collectively bent on succeeding; David versus an industry Goliath. But that was a singular story. What Mary Jane created was a plethora of success stories about Kronos customers using the overall company history as each tale's backbone. The stories were customers, while Kronos was the backroom engine that enabled their financial improvements. Readers devoured accounts about Sears or Marriott or any of tens and then hundreds of companies they knew and cared about, while learning that their secret ingredient was a partnership with a small, growing company in Massachusetts.

"I would never directly or overtly promote the company," Mary Jane divulged as her key to getting positive placement in key publications. "The release would contain what the customers said and felt. Sometimes the magazines would write a story themselves using our release as background. But often, particularly in trade publications where many didn't have the staff to do major rewrites, we made it easy for them to cut and paste. And they wouldn't object because we were not overly aggressive in promoting ourselves. We let the results speak for themselves.

"Then," she continued, "we'd reprint those articles as testimonials for our sales force to utilize. Because hearing good things from other people is always better than hearing it from sales."

This ongoing process continued as Kronos grew and attracted more and bigger customers from ever-more diverse industries. Most articles appeared and promoted the company according to plan. But, in looking back, both Mark and Mary Jane immediately identified the same glaring exception.

"*Forbes*," Mark said succinctly of the influential American business magazine known for many annual lists of "Bests" or "Top."

"They were with us for over an hour and a half," recalled Mary Jane. "And when they wrapped up, I remember Mark saying 'Mary Jane . . . I have a really bad feeling about this.'"

"It's wasn't anything terrible that happened during the interview," recalled Mark. "But I just didn't feel good about it."

By then Kronos had been featured in trade magazines and newspapers. It had received positive reviews in the prestigious *New England Business Journal* and even *Fortune* magazine. And it wasn't that the article didn't open with a bona fide success story about a company of 12,000 deploying Kronos clocks with great results, or follow-up with the backstory of an industry rattling company. The issue was the headline, cherrypicked from the last third of the article.

"Why do technology pioneers have trouble running companies?" the kicker above the main headline asked. "Consider Mark Ain, inventor of the electronic time clock."

And then came the main headline . . . in a font three times larger, in boldface, and with quotation marks: "I'm a bad manager."

No analysis of how Mark had protected his investors and, in so doing, controlling interest in the company . . . nothing about how Kronos, with Mark at the helm, had continuously poured potential profits into R&D so that now his clocks read mag-striped badges . . . two key factors that would serve as key moves in establishing Kronos as more than another of the technology companies of the day that would not see anything close to a forty-plus year run.

"Mark Ain's problem was that he was too concerned with growth to think about profits," the *Forbes* reporter wrote.

"Yeah . . . I'm that. I'm a bad manager, right?" Mark could chuckle decades later, after being named both New England Technology CEO of the Year and Entrepreneur of the Year by the Entrepreneur of the Year Institute.

"They took that one little statement but left out the larger point Mark was making," said Mary Jane. "He said that at that stage of the company's development, he had realized what his role needed to be. He had to surround himself with the right people so that, as a team, they could continue moving the company forward."

"That's a downfall for a company when their CEO refuses to take a hard look at himself or herself, and a hard look at the people around

him or her, and realizes that he has to make some changes," Mark
recounted. "That's what I said in the interview. But that didn't make it
into the article."

"It was our first uncomfortable interview and our first article where
we didn't like the results," said Mary Jane, who recalled that one influential investor had gone so far as to write an open letter to the editor saying the article as written was not factually true—what today we
would call "fake news."

"But don't forget," Mary Jane quickly added, "you're still in *Forbes*
magazine! So, you go into spin and damage control after that headline. The rest of the story was saying that this is a successful company
. . . and the message is out there that the solutions come with an excellent ROI . . . in *Forbes* magazine!"

Or, as one Phineas T. Barnum often said, "There is no such thing
as bad publicity!"

Mary Jane's impact at Kronos would expand far beyond securing
publicity as she soon added investor relations to her burgeoning list of
responsibilities as the company prepared to finally go public. In that
role, she essentially became Mark's coach.

"In the beginning of this process Mary Jane really carried us," said
Mark of this monumental achievement. "I was not very good at public speaking at the time because I didn't do a lot of it. So, Mary Jane
put together an outline for how we should present the company to
potential IPO investors. And she and CFO Paul Lacy and I would travel
all over the United States. We even went to Europe for a week seeing
investors."

"It was exhausting," said Conary. "The preparation before meeting
with the lawyers . . . and putting together the prospectus. None of us
had done this before! But we did have fun during those rehearsals . . .
even though Mark didn't like to rehearse very much."

Several years later, after ringing the bell at Nasdaq, Mary Jane continued the push, serving as the initial point of contact for all investor
related inquiries, and Kronos, based on a solid communications plan
and what would become a historic run of consecutive quarters of revenue and profitability growth.

"It was hard work," Mary Jane would look back. "But it was also very
rewarding work because of the level of success the company enjoyed."

Along the way, and during a tenure that included many high
points, Mary Jane even did something most within Kronos would have
thought impossible: she convinced perennially tight-fisted Mark to
spurge on a special treat for top-performing employees.

"You know," she said, "Mark really watched his dollars and cents, which was part of our financial strength and, more than that, a solid part of our culture. But I went to him and suggested we create an incentive program where we'd go all out to entice the sales force to never, ever give up on a deal."

The program, which became known as Legend Makers, was essentially an annual trip of a lifetime that was bestowed on salespeople who had exceeded quota. The reward was the carrot. But the true destination was ongoing success for Kronos.

"It was a lot of money," said Mark of Mary Jane's incentive program. "But it was really one of the best things we ever did."

Almost as good as the decision to hire a junior ad executive as his director of corporate communications and marketing communications after only a few interactions, because, Oscar Wilde-ly enough, the move truly got people talking . . . about Kronos.

Chapter Thirteen

FIRED UP

There's nothing wrong with being fired.

—Ted Turner

As Paul Lacy waited in the reception area of Kronos headquarters adjacent to what was then proclaimed as "America's Technology Highway" in Waltham, Massachusetts, he kept repeating one phrase over and over in his head as he awaited his initial interview for the position of Chief Financial Officer.

"It was a hot, hot, hot day," Paul recalled of that day in the summer of 1988, "and the air conditioner was broken. I was sitting there, sweltering, waiting for a 4 p.m. meeting as the clock clicked to 4:20 p.m., and then 4:25 p.m., and I remember repeating to myself over and over again, 'Just say no! Just say no!'"

Paul sat there in an oddly enviable position. He wasn't seeking a job, but neither did he have one. He didn't have a current position because he'd been abruptly fired from his last job, yet he'd been let go for a near-perfect reason.

"I had recently been fired as the CFO of a start-up in Canada," Paul recounted, "because I refused to book revenue that wasn't good revenue. The CEO of that company had told me to book revenue that had come in after a quarter had ended. And I had told him I wouldn't do it."

That Paul had a legal understanding of the ramifications of inflating the books was apparent. He did, after all, have a relatively unique background in that he was both a CPA and an attorney. But well beyond the legal side of things, what he was being directed to do was not ethical. And, Paul, as would be a hallmark of his career, held course with a moral compass that never deviated from true north.

"The CEO of that former company said we'd be fine. Just do it!" said Paul. "And I said I wouldn't. He told me if you're not going to do it then you're fired. So, I was fired."

Fired, but with a severance package that would pay him two years' worth of salary. Paul would essentially be allowed to spend quality time at home with his wife and, at the time, two young children.

Some banking acquaintances who happened to have provided a loan to Kronos, knowing Paul's background, pushed for him to at least go to Waltham. And then a phone call from Mark Ain closed at least that part of the deal. Paul agreed to visit headquarters.

"But I kept saying to myself," Paul reiterated, "I'm really not interested . . . just say no!"

But that day, with sweat on his brow, and elsewhere, while speaking with Mark and others, there was a revelation.

"Something just clicked," Paul laughed. "All of a sudden instead of no, I'm thinking 'I can do this!' Kronos had great leadership. They had a great idea. They were basically creating an industry. And even though I went in thinking the opposite, I was in. It was a good match from the very beginning. And it only got better."

"Paul had all the qualifications for the job," said Mark. "And he and I immediately sensed a fit."

In a deviation from many of Mark's principle hires to that point, the "fit" he and Paul both sensed so quickly was one of complimentary approaches rather than matching styles.

"We were very much opposite," detailed Paul. "I could tell right away that Mark was a risk-taker . . . a damn the torpedoes, full-speed-ahead type. He could come to a conclusion, and often the right one, very quickly through a mix of input and intuition. I'm much more data driven. I want to see the numbers. But you need people with different skills and different viewpoints to make a winning team."

It would be about a year before former Los Angeles Laker Girl turned singing sensation Paula Abdul would rise to the top of the pop charts, singing about how "Opposites Attract." But where Mark, Paul, and Kronos were concerned, the song simply hit the turntable a little late. Bolstered by a solid dose of mutual respect, the two meshed together almost instantaneously.

"Right out of the box came a situation that proved to me I'd made the right decision to ignore my earliest misgivings," Paul remembered. "Right after I joined Kronos came my first end to Kronos' fiscal year, which was on September 30. And when we say the fiscal year ends on September 30, we mean midnight on September 30."

Year end is a hectic time. Orders get written at the last minute. Product is getting frantically shipped at light speed. Chaos is an understatement. And amid that backdrop, the Chief Operating Officer at the time, who oversaw the manufacturing of the Kronos time clocks, kept shipping right past midnight and steamed through the following morning, all the while counting those Cinderella clocks toward revenue for the just completed fiscal quota.

"So here I am," said Paul, "the new guy on the block, and I said to him, 'that's not revenue.' And, of course, he disagreed pretty strongly. So, he took it to Mark. And Mark backed me up. He told him that was my call!"

That COO didn't last much longer. But Paul had jumped on board for the long haul.

"I became Employee number 135. And I still have that badge to prove it," he said proudly.

For the next twenty-plus years, as both Mark and Paul would agree, Kronos, and its management team, matured . . . and prospered . . . because good decisions were made, and the company continued to attract hard-working, intelligent, honest, fun people. But in that immediate stretch beginning as soon as Paul said yes to the opportunity he'd vowed to decline, there were immediate priorities in the wheelhouse.

Right out of the gate, there was a need to bring order to the executive team so that decisions could be balanced for the organization as a whole. Acquisitions would be another priority, as it had become clear that the increasing depth and complexity of Kronos' emerging product line required a highly trained, highly motivated sales force to complement and financially drive the ongoing investment in engineering. There was also a building sense that Kronos, in a manner similar to *The Little Engine That Could*, should focus on a journey toward an eventual IPO. All paths pointed toward growth, but each avenue would require focus and commitment.

"I won't say it was like the Wild West," recalled Paul of that late 1980s period. "But the company was a lot less sophisticated than it would be. And from a management perspective, it seemed like a lot of decisions were made by whoever yelled the loudest."

The need for each contributor to assert his or her alpha status wasn't necessarily personal. As would be learned more fully later during an assessment exercise, most within the core group were somewhat predisposed to taking definitive and sometimes aggressive stances.

Recognizing that fact let some air out of the grossly overinflated balloon in what was then the boardroom. And adding Paul, who again was more of data-driven, less emotional, and more logical thinker to the mix likewise brought another tension-easing voice into the conversations. But Paul also readily gave credit to Mark's leadership skills, as well as his innate ability to understand what not only wasn't working, but what had also started to turn toward the realm of counterproductivity.

"Mark realized early on that this type of atmosphere wasn't healthy," recalled Paul. "He was very good about asking people's opinion and incorporating those opinions into a decision in a way that allowed mutual respect to grow."

That didn't mean everyone fit perfectly into a company mold. Though a majority of the early leaders stayed put, some moved on, perhaps because they sought other opportunities within a different environment, while others were given more direct encouragement to leave.

"As we matured," said Paul, "most people realized that the yelling and infighting wasn't a healthy process."

What Mark observed was that Paul was a positive catalyst in that somewhat overdue maturation in the board room. And that caused him to revise whom he sought out for consultation at certain points. In that regard, Mark's interpretation of Paul's strengths became a long-running tradition that would benefit both Kronos and Mark's style of management moving forward.

"I recognized that beyond the board room, Paul's profile was more like the profile of the average employee within the company," said Mark. "So, many times before I would make a decision, I would seek out Paul to get his perspective."

"Mark had an innate sense that he needed to hear from people who had a different perspective," confirmed Paul. "And I think that's something very rare for a CEO. Most CEOs, it's their way or the highway. You agreed with them or you were gone."

Kronos, having become profitable some years earlier, continued to grow both in revenue and profitability, while also maintaining Mark's longstanding policy of pouring significant financial resources into engineering. And with Paul's steady hand on the inner workings of the company, with him at the reins of credit and collections, legal, financial reporting, and HR, Mark was further freed to keep pushing the company to higher levels, safe in the knowledge that Kronos was likewise run as well as the companies Kronos sought to assist in the managing of their internal operations.

"The wheels on the bus had to keep turning," Paul said with a chuckle. "My role was not as sexy as getting new products to market or adding new territories. But I relished my role. And, as time went by and I came to understand more about the business we were in, I was able to provide direction on how we could better run our operations."

To that end, Paul was grateful that most of the folks with whom he dealt internally were, like Mark, receptive to his input and perspective. Again, he saw this as a harbinger of greater things to come.

"Other people would listen to me. I, in turn, was willing to listen to other people," he said. "We didn't always agree with each other. But I can say that I worked closely with Mark and his brother Aron and most everyone else, and, even when we would disagree, there was never a cross word between us."

To illustrate that he took input as well as he gave it, Paul remembered a time when the current head of the legal team was stepping down. Paul had begun a search for a replacement when Mark came to him.

"Mark came to me and basically preempted the decision-making process," he recounted.

By this point, Kronos had become a sizeable company. Naturally, Paul was seeking a candidate with experience at that level. Mark, on the other hand, felt that a potential internal candidate, though a relative junior level attorney, was the right fit from both a personality and domain knowledge perspective.

"I wouldn't have thought of her," Paul confessed. "But Mark was always great at seeing both talent and potential. He walked in and said to me, 'I think we ought to give this to Alyce Moore.' And you know what? He was right! She stepped up and ran that department as we grew from about $50 million in annual revenue right up into a multi-billion-dollar organization and never missed a beat!"

Paul felt a move like that was indicative of Mark's less analytical but ultimately positive nature.

"Mark seemed to always have this gut feel for who was the right fit at the right time," said Paul. "Whether it was in manufacturing or service or anywhere, he knew what he had to do and did it. But bigger than that, Mark has a very good heart. If someone outgrew their position, or even if a position outgrew someone, he'd say 'Maybe there's another fit for that person.' If you were a loyal, good employee, he would do what he could for you. And he possessed this sixth sense for when a person was right for a job, and even when they were not.

"Because he had a good heart," said Paul, "that's the type of company he built . . . one made up of other people with good hearts."

Simultaneous to refocusing the board and overseeing internal operations in coordination with Pat Decker and Aron Ain, Paul was also heavily involved in dealer acquisitions, as Kronos began the process of evaluating, purchasing and absorbing the dealerships that had originally comprised the Kronos external sales force.

"From the time I came in and over a period of about 10 years, we steamrolled through the process, ultimately resulting in fifty-seven transactions," said Paul of Kronos' transition to an internal sales structure.

"Paul's role in dealer acquisitions was critical," said Mark. "We had a specific formula for acquisitions. And it was very detail oriented. It wasn't perfect. And from time to time we would deviate from it to bring the right dealer on at the right time. So, Paul, with his combination of skills, was a major part of that process as well."

With the Kronos organization continuing to evolve, and with ongoing revenue growth as a driver, Mark and the management team set sights on what had come to represent a major milestone for any of the successful technology companies of that era: going public with an IPO.

As CFO, Paul would have a major role in this sometimes complicated, often arduous process.

"It was daunting," said Mark. "And it would never have been accomplished without the efforts of a lot of people. But Paul's contribution, like Mary Jane Conary's, was critical."

"As one of the companies during that time, especially where we were located on America's Technology Highway, going the IPO route was a rite of passage," reported Paul. "At that point we didn't have a lot of venture capital money in the mix. But there was a promise that those investors would get a reward for their investment. We had other debt to take off the books. And we were looking for capital to fund additional ventures, as we looked to grow through acquisitions and investments in the technology."

Following the IPO route also meant quite a bit of travel. The core team of Mark, Paul, and Mary Jane traveled in and out of the country, putting on road shows that detailed the current and future outlook for Kronos for potential investors.

"You had to deliver presentations and then answer some very probing questions," said Paul. "We were asked to do a lot of face-to-face meetings. So, there was a lot of go, go, go! Without question, we got to know each other even more through the process. So, it was good that we enjoyed each other's company."

Investment bankers lined up to represent the company at IPO. But even after the bell rang on Wall Street in 1992, the pressure was nonstop to present and represent.

"We'd raised $18 to $19 million, which was one of the smallest IPOs at the time," said Paul. "It positioned us to grow the company. But it also meant that at least every quarter, one or all of us would have to travel around talking about the just-completed quarter and laying out future plans."

"It seemed like after the IPO we were constantly traveling," said Mark. "Every quarter we packed the suitcases and off we went."

"The upside though," added Paul, "was that we were making Kronos more well known, which was a good thing. So, we were adding investors while also improving the perception of the company."

"I recall that at the time of the IPO we were over $50 million in annual revenue," said Mark. "After that, we just grew like a weed."

"Our overall philosophy," said Paul, "was not only to invest to become a bigger and better business . . . but to create a marketplace . . . to build the business while also building an industry."

Kronos had already begun to expand beyond time clocks, which in and of themselves had really become more like mini-computers with ever-expanding capabilities. But by expanding the vision, Kronos began to branch more heavily into software-based solutions and offer customers greater control over payroll and scheduling and HR functions, eventually redefining the industry as Human Capital Management.

"We didn't add dividends. And we didn't initiate any stock buy backs," said Paul. "We just kept pouring the resources back into the company."

"Paul and I were the yin and the yang of it," chuckled Mark. "We discussed every move because his perspective balanced mine. Our relationship was really a partnership."

In fact, Mark came to see Paul so often that when the company moved its headquarters from Waltham to Chelmsford, Massachusetts (company HQ is now in Lowell, Massachusetts), Paul requisitioned to have a couch in his office.

"He approved me getting the couch after I explained that it was mostly for him!" Paul quipped. "Mark would just walk in, sit on the sofa, and we would just talk."

"By talking to Paul, I could get a true understanding for how our employees were feeling, or would feel, about a move we were planning," said Mark. "And that was the balance I needed to keep running the company for as long as I did."

The track record of success all but speaks for itself. But, as for the guy who sat in the reception area in sweltering heat telling himself over and over again to turn down Mark and Kronos, Paul is glad he kept his cool.

"Being CFO at Kronos . . . it was my dream job," he stated emphatically. "The time went so fast. We had our challenges . . . like 9/11. But we weathered through because we really had the dream team, a right mix of personalities that contributed to a long series of successes due to hard work by good, smart people . . . people I was truly pleased to work with."

Sounds like a guy who was truly fired up!

Chapter Fourteen

THE "PLASTICS" OF THE 1980s

You can't build a reputation on what you're going to do.

—Confucius

Not that every seasonal shift brings about dramatic new beginnings and change, but the spring of 1981 held significant impact for both the immediate and long-term trajectory of the world.

Amid a major economic downturn—with inflation at 10.35 percent and home mortgage rates cresting at over 18 percent—Ronald Reagan became president, 3M launched Post-it notes, and the first (and soon to be last) stainless-steel DeLorean cars took to the roadways of America.

The first Space Shuttle—*Columbia*—blasted off for the thermospheric layer of the atmosphere, Iran released its American hostages after 444 days, Mohammed Ali retired, and a heretofore relatively unknown woman named Lady Diana Spencer married longtime bachelor Prince Charles of England.

Yet, within that turbulent and troublesome time, recent college grad Jim "Kizzy" Kizielewicz was thinking about a movie from 1967 starring a young Dustin Hoffman. The movie was titled *The Graduate*.

In the film, Hoffman plays Benjamin Braddock, a twenty-one-year-old recent college graduate himself, as he looks to find direction in his life. The central plot in the award-winning flick revolved around Benjamin having an affair with the much older wife of his father's business partner, but Kizzy, the newly minted Hamilton College alum, was pondering a scene in which another neighbor, identified only as Mr. McGuire, seeks Benjamin out to provide some career advice.

"I want to say just one word to you," McGuire says to Benjamin in hushed tones.

"Yes, sir . . . " replies Benjamin.

"Are you listening?" McGuire insists.

"Yes, I am," says Benjamin.

And McGuire leans in as if delivering the launch codes for a nuclear weapon.

"Plastics," he says before nodding knowingly.
"Exactly how do you mean?" asks a slightly befuddled Benjamin.
"There's a great future in plastics," nods Mr. McGuire. "Think about it. Will you think about it?"

Benjamin agrees to think about plastics. And Mr. McGuire struts away, convinced he's set the young man on a path to prosperity.

In 1967, plastics represented a bold new frontier. Nowhere near a time when there would be a literal island of plastic garbage roughly the size of a smallish continent floating around atop the ocean, instead, plastics in 1967 were poised, at least in a corporate sense, to explode. So, finding a job that involved plastics could indeed launch a career.

"That scene in *The Graduate* where the kid gets out of college and finds himself at some upscale cocktail party and this guy pulls him aside says 'plastics . . . get into plastics' had stuck with me," Kizzy divulged.

With Dustin Hoffman swimming in his head, Kizzy pondered the landscape that was 1981 and saw a parallel potential within the advent of computers. To Kizzy, computers, not yet running the world but seemingly gobbling it up one gigabyte at a time, represented his future. And given the uncertain economic climate, and the fact that he had student loans that would need paying, he was willing to get dirty to establish a foothold.

Truth was, before graduation, Kizzy knew very little about Kronos . . . or about computers. And he knew somewhat less about job seeking, having graduated without having yet crafted a resume with which to go hunting for a regular payday. What he did know was that good friend and Hamilton fraternity brother Aron Ain had found a job the previous spring at some company up in the Boston area where he worked for his older brother Mark, who had founded a company based on the premise of putting tiny computers inside a time clock. And there was a tactile connection. Kizzy was thinking computers. And his father had spent his working life punching a time clock.

"Looking back on it," he said, "I was thinking the equivalent of that scene in *The Graduate*, in 1981, was computers. I wanted to get into computers because I knew it was going to be a growth industry."

So, with no true job prospects that aligned his degree in government with his forward-thinking vision, Kizzy reached out and asked Aron how he was doing. Aron replied by indicating he was happy with his choice, and that a road trip north might be a good idea.

"We were in the midst of a recession and I was basically trying to decide what to do after Hamilton," he recalled. "I had heard that Aron's brother . . . I hadn't even met Mark at that time . . . had started a company. So, I got in touch with Aron, and he said there might be a role there. To make a long story short, I made the trip to Boston."

Meeting with Aron and some other Kronos folks earned him a chance to meet with Mark the following week. And Mark once again looked past the resume (or the lack thereof) and saw someone who would fit on the team. The question right away was where?

"We didn't really have anything at the time," said Mark. "But I liked him. So, I told him we had some openings in shipping."

"At first," said Kizzy, "Mark told me I'd likely be in shipping for about six months. Then we met again and Mark said they were now only going to keep me in shipping for three months."

A job . . . and as important at the time, a salary from which bills could be paid. And a direct connection to his "Graduate" theory.

"Kronos was going to be built around a computerized time clock, but a computer nonetheless," he said. "So, what really appealed to me about Kronos was that it was getting in on the ground floor of this new industry.

"And," he continued, "during our talk Mark had told me that one day Kronos would be a $100-million-a-year business. And I was, like, wow! I don't know how we're going to do that. But I'm in!"

As it turned out, Kizzy didn't have to start at the ground level after all, though he only learned that the night before his first day on the job while unpacking in the apartment he would be sharing with Aron.

"The day before I was to start, I'm going through my clothes to find some jeans and a shirt I wouldn't mind getting dirty because I'd assumed you'd get really dirty in shipping," he said. "And Aron comes and tells me I'm not starting in shipping after all. They'd decided to move me right into what was then called Marketing Services."

The pre-employment promotion meant slightly better clothes, and the chance to really learn the business.

"Back then the Marketing Services group . . . we got everything. There was a group of about four or five of us," said Kizzy. "We basically did quality assurance of the product for engineering. We wrote documentation for it. We helped the resellers program. We handled tech support questions when there was a problem. We wrote up sales orders. We trained the dealers on how to configure the system. Then, at the end of the month we would go down and basically help manufacturing to build the clocks."

Everything just short of shipping . . .

"We did whatever had to be done to support the business," he said, laughing. "I can't speak to the quality of the builds when we went down to manufacturing to help. But we got them out the door!"

And did he mind the jack-of-all-trades atmosphere?

"Not at all," Kizzy stated emphatically. "I felt like somebody was giving me a chance to prove what I could do at a time when, with the recession, jobs were hard to find. I didn't have any computer experience. I wasn't an engineer. But here I was . . . given an opportunity to try and prove myself and really make a difference. So, from my perspective, it was a great learning experience being able to do a lot of different things."

Armed with a working knowledge of what the early clocks could and, perhaps more importantly, couldn't do, Kizzy kept helping wherever needed, but soon began to find a strength.

"Over time, I began to gravitate toward the product side of the business," he recounted. "In those early days we were really only shipping one product—Timekeeper. That device was selling for around $5,000. And it was doing well. But the market was limited. And we were getting pushed, really, in two directions. Some people were asking for more capabilities, whether from us or someone else. And we had a lot of people asking us to build a cheaper model."

"Kizzy had a really strong sense for where we needed to go with the product," said Mark. "I knew right away that we had to continuously innovate and respond to customer needs and desires. Kizzy got that. And he was involved in every generation of product advances from the very beginning."

"Listening to our customers really set us on a path to build a suite of products," Kizzy said.

"We had an ongoing commitment to get better," said Mark. "And I took some early criticism for that. Because rather than take whatever profit we could, we stayed committed to reinvesting in the company. To a large extent, reinvesting meant constantly reimagining the product. It meant spending money on making the current product better but with a sharp focus on expanding what we could offer to our customer base."

Kizzy was in lockstep with Mark in his belief in getting better. He'd felt before meeting Mark that the transistor technology that begat the advent of computers was the future. But now that he was on the inside, at least within the realm of timekeeping, he could clearly see that the

collective effect those devices had within the business world was growing exponentially.

And, as Mark and his earliest collaborators had believed when
they were fashioning prototypes for the original clock and housing
them inside old printer casings, the relative cost of materials, due to
expanding production, was dropping, the result of which was that,
once designed, tested, and ready for the market, newer solutions with
greater capabilities built in were becoming less expensive to build.

Beyond cost reductions, what was about to happen within Greater
Boston's technology corridor was revolutionary. The true fast lane was
being built with PCs—personal computers.

"Customer input put us on the path to build a suite of products, or
multiple devices," recalled Kizzy. "And the costs going down allowed us
to break a price point . . . now our products could be sold for less . . .
and that opened up a whole new market for us."

Mark had established Kronos by identifying an industry led by a
large player that was set in its ways and unwilling or unable to adapt
to an evolving marketplace. Now that Kronos was showing signs that
it could (and would) take over the time clock space, it was not time
to pop champagne corks . . . it was time to innovate and deepen customer relationships by making a more viable and further reaching
series of solutions available. And while the team Mark had assembled
all but unanimously embraced that notion, in Kizzy he'd found a driving force who shared his vision.

"I think if you look back over Kronos' history you'll see that the
one thing we've done, over and over and over again," said Kizzy, "is
this concept of recognizing that the market, or the technology, is shift-
ing. And we consistently shifted our platform, and our solutions, to the
market and the technology."

Due to that level of commitment . . . that don't-rest-on-your-laurels
mentality, Kronos got out in front of the technology wave that was rolling in and inundating all things business. But the burgeoning landscape came with an early trap. A plethora of many different companies
were out there inventing technology, and capabilities and capacities
were growing by leaps and bounds. But most of these companies were
operating in vacuums. More than that, they were very reluctant to
share their innovations on a broad scale for fear of losing the proprietary advantage that could allow them to claim superiority over one
another. So, for a company like Kronos, one that was adjusting its now
growing suite of solutions to these many technological advances, there
came a question of, not *how* to align, but with whom . . . because by

then the genie was out of the bottle. Kronos customers and prospects alike were not just buying time management systems. They were buying computer systems first, and then expecting that Kronos would ensure that its products would align with the massive investment they had made in a given computer system.

Early on it wasn't so much about who the Kronos team might *want* to align with, whether it be Digital or Apollo or Data General or Wang or any of the key players at the time; it was more about how to integrate with whatever room full of stacked equipment was housed in a given company's ventilated and air-conditioned information technology room. And doing that . . . was not easy!

"We were trying to figure out how to integrate with all those different platforms," recalled Kizzy, "and I'll admit, it was a bit of a nightmare."

A nightmare worth attacking, and with good precedent. After all, the world was changing in ever-shortening cycles as technological advance begat technological advance. And if the earliest days of Kronos had brought a lasting lesson, it was that standing pat as the alpha dog was akin to being a mighty dinosaur standing tall as the tar oozed around and began to engulf its mighty feet. The future lay in constant movement both forward and toward higher ground. Constant adoption and reinvention could not be an occasional deviation from the day to day. Evolution *was* the day to day!

Kizzy's role, and impact, expanded in concert with the capabilities driven by technology. He was not an engineer, but his vision made him a driving force for change.

"We started within an industry that was sleeping at the wheel," Kizzy stated. "And we cracked that mechanical time clock industry by disrupting the incumbent player. We saw how Kronos could come in almost unnoticed and so we were committed to not becoming that company that didn't see change coming. The thought process was to imagine ourselves as that little start-up company that wanted to get ahead of us. Technology can change from underneath you if you don't approach each day with a sense of paranoia. So, we reinvented ourselves over and over and over again."

So, as the dust settled around the collapses of many an early technology company, there was Kronos embracing changes like the advent of the PC, and the connectivity of the internet. Always mindful that there were upstarts trying to make headway in one niche or another, while at the same time never losing sight of the fact that the solutions had to be reliable first and innovative second, Kizzy and Mark and

Kronos' whole team acted simultaneously like an industry leader and a start-up.

"There was this constant feeling that we had to go out and disrupt the incumbent player even though we were the incumbent player," said Kizzy. "That paranoia stuck with us, and drove us. Why wait for some other company to disrupt us? Let's disrupt ourselves."

Which isn't to say that other organizations didn't eat market share. But within Kronos, Mark had chosen leaders, visionaries, really, who could anticipate and embrace change. And though many potential disruptors would sometimes try to lure certain of those leaders away, a preponderance of those early employees, Kizzy among them, mainly opted to stay in the fold.

"Sure," Kizzy, who would rise to the role of chief marketing officer at Kronos and the chief operating officer of UKG, reminisced, "every now and again I'd hear from some other company about maybe looking elsewhere. But I never took it too seriously because Kronos, with its market leadership and so many great people already there, was a great place to work. And besides, because Mark had hired me right out of college, and almost before I got the chance to seriously start looking for a job, it would have been hard for me to look elsewhere anyway."

And the reason it would be hard?

"The timing being what it was," he said. "I never got the chance to write up a resume. I think I wrote something up in high school when I was looking for a summer job. But it was never updated."

Besides, once Kizzy had found his own version of plastics, why look anywhere else?

Chapter Fifteen

GROWTH WAS IN THE CARDS

Each player must accept the cards life deals him or her; but once
they are in hand, he or she alone must decide how to play the cards
in order to win the game.

—Voltaire

Two letters and one numeral led the business news cycle in the
two- to three-year stretch leading up to the ball dropping in Times
Square to signal the beginning of the earth's third calendarial millennium . . . Y2K.

Predictions for horrific disasters of all scales ran rampant, based
on the notion that all the technologically driven innovation of the previous decade would shut down based on a simple premise . . . computers and anything driven by them would be rendered unworkable
because all these gadgets marked years with only two, not four, digits.
Presumably, all these machines would read double zero on their internal clocks at the stroke of midnight and, "thinking" it was 1800 or 1900
and that they had not been invented, they would shut down.

Airplane guidance systems would fail, with planes crashing around
the globe in sequence to respective time zones. Banks would crash as
well, with no monetary tracking. And worse, Kronos time clocks would
not tabulate hours worked!

The air of potential horror permeated everywhere as software
experts around the globe sought to claim that their systems were
Y2K-compliant.

Interestingly, and despite how ridiculous this sounds so many years
later now knowing that these Nostradamus-esque forecasts bore no
fruit, the calamity did cause a financial wrinkle in Kronos' plans. Tech
companies had been going gangbusters, driven by the dot-com boom
and what was ultimately a Y2K version of "fake news." Organizations
the world over had ramped up tech spending to thwart the forecast of
calamity. The ball dropped in Times Square and nothing popped but
champagne corks, while the monies essentially borrowed from future

budgets left many wells near dry entering the new millennium. Coupled with that drain came a downturn in the economy as dot-com after dot-com, buoyed initially by promises of great and wonderous things blossoming with the advent of the Internet, went belly-up.

It could have been life-sapping at the young Kronos but for what could be termed a paper-thin safety net . . . time cards!

Early Kronos time clocks had been accepted over what was likely a better offering from Simplex because Mark's research had told him that U.S. employees, wary of technological advances in the workplace, would not readily embrace the clocks unless a time card was involved. So those initial designs included a patented method for the reading of paper cards—paper cards that Kronos supplied.

So, at a time when securing new orders became tougher to come by, Kronos was able to generate a revenue stream selling clocks into a recession-proof nursing-home industry and supplying current customers with employee time cards.

"These were specialized time cards that the clocks would read when they were inserted," said Mark. "But they had to be replaced by the boxload. So, we could forecast, based on the number of employees a given customer employed, a relatively constant revenue stream for as long as those customers had the clocks that required time cards."

Being able to forecast this revenue provided Mark with an opportunity to somewhat invoke his company's namesake—Kronos, one of the Titans of Greek mythology, the Greek god of time. Kronos was not necessarily father of the year. Fearing any one of his children might overthrow his rule, he ate each of them whole upon birth. Zeus was the lone exception, but only because Kronos' wife Rhea tricked him by putting a large rock inside a swaddling blanket when Zeus was born. Kronos gulped down the boulder, thought his latest offspring was a little tough, and went on with his godly day. Zeus, raised in seclusion, was eventually able to force Kronos to regurgitate the others. Since they had been swallowed whole, they were all fine.

Perhaps not consciously following said Greek mythology, Mark still followed suit as he began to reinvest this regularly scheduled revenue by systematically devouring his own sales network whole, in favor of regurgitating it back out in the form of an internal sales force.

To that point, the sales network had been mainly populated by outside dealers who sold the Kronos time clocks on a regional scale against a quota system that rewarded success. Kronos had established its own regional sales offices in places like Chicago, Los Angeles, and Atlanta, but those were the outliers. Now, with traction growing under the Kronos revenue stream, Mark saw an opportunity born from necessity—not an opportunity to cash out, but rather to reinvest and point toward a much longer-term horizon.

"By the year 2000, we'd gone from start-up to the 800-pound gorilla in Time and Attendance," said Mark. "Early on, the only city we set up direct sales in was Chicago. Then we'd added LA and Atlanta. But a bulk of our sales came from dealers where we set up what was, at the time, a very unique way of setting a quota where, if a dealer cleared a quarterly quota of sales, they got a discount on everything they sold. So, they had great incentive to really push the product. The bigger the quarter, the bigger the discount. It was really revolutionary."

But, while it was a revolutionary way to incentivize dealers to push hard for Kronos at the time, by 2000 it was time to change up the paradigm.

"By 2000, many of the dealers were falling behind," Mark recounted, "either because they didn't have the money to invest in their own infrastructure, or they didn't have the technology smarts to keep up with us."

The infrastructure issues definitely stemmed from the economic slowdown. But the technology curve was an even bigger deal. And Mark could readily see that Kronos' internal offices were savvier to the updates and releases specifically because they were staffed with Kronos employees.

It was after this evaluation that Mark knew that it was time to bring most of the sales channel in house. And that meant it was time to recruit someone like Bill Breen to join the corporate offices to help drive acquisitions.

Bill had been working at Fidelity Capital, the arm of financial behemoth Fidelity Investments that invested in various businesses.

"I came on board in February of 2000," Bill said. "I remember that time because just about three weeks after I joined Kronos, on March 8, came the start of the end of the dot-com bull market. Tech companies that had been doing great the prior four years very quickly fell into trouble. Kronos took a hit, but not as hard as most other companies."

Bill was hired as the lone financial analyst charged with handling the front end of deals. Kick-the-tires kind of stuff. Valuations . . . cash flow . . . outstanding debt . . . factors that could be added to a calculation that would determine a price by which Kronos could get a fairly rapid return on investment. At the time, he was thirty-two years old!

"At Fidelity Capital, I was one of thirty people on a team. There were ten of me!" said Bill. "Kronos was 100 times better than I thought when Mark hired me. This was a chance to really be involved."

And involved . . . and involved. In fact, over the course of the next seven years, Kronos would acquire a whopping thirty-five businesses, both small and large, in deals ranging from a $350,000 dealer acquisition all the way up a $151 million merger that was, like most "mergers," an acquisition by another name.

The initial push was to take what was a predominantly outside sales force and transform it into an internal one. By doing so, Kronos would gain greater control over the sales processes from both a fiscal and a quality perspective. This move made great sense because the technology was advancing rapidly, meaning there was an increasingly sharp and constantly evolving learning curve where the technology was concerned. Bringing sales under one umbrella would ensure consistent delivery of deals and implementation alike across the board without regard for region.

At the same time, Mark sought to drive innovation and the expansion of the breadth of the Kronos offering by committing to constant innovation through R&D. But rather than operate in a Kronos-enveloped vacuum, the CEO was aware that there were other organizations out there with marketable solutions in both the immediate time and attendance market as well as in adjacent other aspects of what would today be considered the broader market of workforce management. With its ever more dominant position within its own market, it only made sense to gobble up certain competitors and complimentary organizations whose technology and marketplace positions could, once adapted into the realm of Kronos-based solutions, allow Kronos to grow further and solidify the company as a long-term force.

"Basically," Breen said, "there were four different kinds of deals. In the first instance, we were buying dealers, bringing the sales region under the Kronos sales team. The second was buying competitors, where we were buying their customers and converting them to utilizing Kronos. Third, we bought competitors for their technology, so it could be adapted into or enhance the Kronos offering. And finally, we bought companies that had technology in adjacent technology within workforce management."

While moving into this important expansion phase, Kronos was also finding yet another revenue stream, one that would fill in the gaps when the paper time cards were eventually supplanted by ID badges with magnetic stripes . . . another technological advancement in which Kronos served as a pioneer.

"Another extremely important aspect of our revenue growth was maintenance," reported Mark. "Today everybody does this in one form or another. But we saw this in 2000. We developed a service organization that earned us an 85 percent margin. It became a huge driver in our expansion plans."

Properly capitalized and committed to building for the long run versus simply profit taking, Mark felt that, while Kronos clearly benefited, so did those companies that were ultimately acquired.

"We did not screw those dealers," he stated. "They had made good money to that point as dealers. But it was time for a change and the buyouts were fair. And, quite a few people involved in the deal ultimately came to work for Kronos. So, they got good jobs and we found a lot of great, bright people . . . bright, smart people with valuable domain knowledge in our industry."

Bill detailed the process that was undertaken whether the deal was for less than $1 or $2 million or for tens of millions of dollars.

"Mark and the executive team were, of course, the ultimate decision makers," he said. "But you don't want the decision makers taking on the role of looking under the hood. So, I, who reported to CFO Paul Lacy, would perform the due diligence to determine things like whether a company's customer list is accurate . . . do you buy their liabilities with a stock deal. Eighty to 90 percent of our deals were asset deals, so we avoided liabilities like whether an employee of that company had done something wrong that could cause a problem further down the road."

Another key factor was how long it would take for the company to earn back its investment.

"We had to calculate, based on what we proposed to pay, what would the rate of return be . . . how long did it appear it would take to recoup that money . . . and how much risk was there that things might not work out as planned. You had to assess risk at every possible turn," Bill said.

Acquiring dealers was a relatively low-risk move, whereas acquiring a technology company for its solutions was a tad trickier.

"More often than not, 90 percent of a technology company is its tangible assets," Bill explained. "It's about their products. But you need to look to see how they're constructed as well, because every company that wants to be bought trims overhead and tries to boost sales. So, for instance, some dealers, knowing they want to sell, would discount things like maintenance to get more revenue on the books. But later, after they were acquired and it was time for renewal, those customers who'd received deeper than normal discounts would look and say 'Wait, you're doubling the price!' But in reality, maintenance had been sold to them at 50 percent off. You had to look deep to make sure the current revenue was sustainable before you set a price for an offer."

Because all of this was so new to Kronos and, to an extent, Bill as well, the team learned some of the tricks year-over-year.

"We kept adding and adding to our standard Purchase and Sale agreement," Bill said. "In 2000, our P&S was about fifteen pages long. By 2008, it was over forty pages because we found that we needed to keep adding to it to protect Kronos' interests."

Early in this stretch of growth came a major milestone for Mark and Kronos. In 2001, Kronos purchased the original 800-pound gorilla in the Time and Attendance space when the deal was struck to buy Simplex's Time Division, by then a shadow of what had been the world's dominant player in the industry, from Tyco Electronics for $24 million, thus proving one of Mark's initial tenets for success in business: find the industry giant who was behind the technology curve and slip into the industry with a better mouse trap.

Of course, the Simplex deal was not done just to prove Mark's theory. Like all the deals, the objective was to buy for the right reasons and at the right price. In that deal, Kronos bought customers, some technology, brought many of its employees on board, and also eliminated yet another competitor . . .

"That was a big day," Mark chuckled, "But it was also a very good day for a lot of people . . . not just me!"

Mark and Bill would agree that not all deals were letter-perfect. The biggest of the bunch in this historic stretch, the $151 million merger with Unicru in 2006, was undertaken to streamline Kronos' entry into the human resources realm. And while over time the move proved fruitful, first, in bringing Kronos into the HR realm and, later, in solidifying its place as the top workforce management solution provider in the world, it came with many unanticipated growing pains in spite of its massive capital outlay.

"Unicru had products where you could test prospective employees to determine whether they would be a good fit at a company," said Mark. "It was essentially predictive analytics for hiring. But there were problems with the product that we didn't quite see. And their president was a driving force in their success. So, without him Unicru was not the same company we bought."

"Unicru had $47 to $50 million in recurring revenue," reported Bill. "So, the numbers appeared to be there for recouping the investment. But we didn't quantify how much influence their president had on their sales. He was a dynamic force and an undervalued asset."

Ultimately, the investment paid off. But it took longer than anticipated. And it stood out because of its high price tag.

"As we got bigger, we had to do bigger deals," Mark shrugged. "Over the short term it wasn't as expected. But ultimately it was still a good move that helped us over the long term."

Each and every deal stands out to Bill and Mark in equal measure. But as Bill looked back post-retirement, his focus was less on the particulars and more on the experience.

"At Kronos, my sense of self-worth grew along with my enjoyment within the process," he said. "Again, every decision was ultimately made by Mark. And Mark was not someone who was shy about letting you know if you missed a detail or didn't seem to be working as hard and as fast as he did. But in the end, this job was 100 percent better than I ever thought it would be."

"Bill was another person who might not have, at first, looked like the ideal candidate to take on his role," said Mark. "But he absolutely was the right person!"

"I was not the captain of the ship, and that's okay," said Bill. "I had my hand on the rudder. And I approached every opportunity with one thought in mind and that was that Kronos needed the best deal. Mark drove people hard. But he worked hard, too. He gave people the opportunity. It was up to them to succeed. I was very lucky to be involved for as long as I was there."

And based on the results, it was good that Bill Breen was involved. After all, during his tenure, Kronos did need to eat! It was in the cards . . .

Chapter Sixteen

GO ASK ALICE!

But I don't want to go among mad people.

–Alice, in Lewis Carroll's *Alice in Wonderland*

Learning to drive can be stressful. Two tons of steel propelled by a powerful engine. The ever-present need to be aware not just of yourself but everything and everyone around you can be daunting, the potential to do damage or to be damaged overwhelming. The responsibility looms large enough that you have to take classes and pass, not one, but two tests, and then adhere to a series of early restrictions, before becoming a road warrior.

And so, it was with all of that hanging overhead that a teenaged Alice Ain received driving instruction from a brother older enough that when he graduated from the uber-prestigious Massachusetts Institute of Technology, kindergarten-school-aged Alice told friends her brother had attended "some school called Mitt."

That it was a bonding opportunity was self-evident. That it would establish a precedent for Mark serving as a mentor and later for Alice to turn the tables and teach Big Brother some valuable lessons would seem less obvious.

Mark hadn't been an absentee brother by design so much as it was chronological fate. Pearl and Jack Ain had taken a pause after having three sons in rapid succession. Aron came fourth in line and then a change . . . a baby girl . . . Alice! But with a span of roughly fifteen years between them, Mark was further down life's pipeline than most older siblings, receiving his MBA from Rochester as Alice finished phonics and the second grade.

Yet there was a strong bond, and a good deal of mutual admiration.

"I remember she was always effervescent," recalled Mark. "Even at a young age, and more so later on, she always had a lot of friends."

"I was always very proud of Mark," said Alice. "He was away a lot of the time because he was so much older, so there were not a lot of household dynamics because he didn't live at home. He was always supportive. And I knew he was doing interesting things. But when he would come to visit it was always a big deal to me."

During one particular visit home, Mark surprised Alice with a gift that stands as a testament to how she cherished their relationship.

"One thing I do remember from those much younger days," she recalled, "is he came home and had bought me this little doll. And my friends would ask me about it and I'd tell them it was from my brother who lived in Boston. And they'd ask why he lived in Boston and, remember I was very young, I'd say, 'I'm not sure.' But I still have that doll."

Not to be outdone, and following in a family tradition of excelling academically, after graduating high school Alice was admitted to prestigious Vassar College, by then coeducational but originally one the Seven Sisters, the name given to the first truly elite women's colleges in the country, then and now one of the top liberal arts colleges in the nation.

Post college, Alice took a job in the world of publishing, working as an editorial assistant for the New York City-based *Hudson Review*. But after a year, Mark, knowing talent and caring little for the norm that many companies practice about not hiring family members, made her an offer she opted to not refuse.

"She was living day-to-day in a small apartment," he recounted, "and I came down for business and visited her in her very small apartment in a very expensive part of New York City working for this firm and I said 'Why don't you join our company?'"

Brother Aron had already been at Kronos for about two years, having joined just out of college. And that was working out. Kronos was flush with younger people like herself. Alice said yes!

"I didn't have any business background," she divulged. "So, I didn't really have any idea how a start-up business would operate. Many of us were just living and working day to day, doing a little bit of everything . . . just doing what had to be done . . . whatever that happened to be on a given day."

Mark was not her direct boss. But, for a short stint, he was her de facto landlord, as Alice initially moved in with him before getting a small place in Allston close enough to Kronos to allow her to either walk or ride a bike back and forth from work.

"Ultimately, I did everything there but write code . . . again, things weren't so defined by job title," said Alice. "But I clearly remember my very first assignment because we had this older woman who was the bookkeeper . . . I think she was in her sixties. And after I'd been at

Kronos for about a week, she fell on her walk to work and broke her arm. So, my very first official assignment was sitting in the accounting department writing checks in the big ledger book with carbon paper to make copies."

Ever-ready to contribute wherever needed, Alice nonetheless knew very quickly that the accounting side of the business was not for her.

"There were no spreadsheets there," she said. "All everyone did all day was look at numbers so everything was so quiet. It was so boring. I just knew this was not for me!"

The bookkeeper's arm healed, and Alice moved back into the fold. She worked testing the clocks . . . in sales . . . in technical support . . . and, because she lived close by, she also often had the unglamorous honor of being the on-call person tasked with running to the office in the middle of the night if an alarm was inadvertently tripped.

"I remember one time while I was working in technical support, I got a call from one customer who said the clock was on fire and he wanted to know what he should do," said Alice. "But I didn't know! I told him to unplug the clock and go get some water!"

Also, around that time, and true to those teenage driving lessons, Alice had another funny they-don't-teach-you-this-in-liberal-arts-college moment.

"I will tell one really quick funny story from those early days," said Alice. "And it ties into the theme of everyone doing everything. There were all these investors coming in because Mark was raising money for the company. And there was some important guy coming in from New York and I was designated to pick him up from the airport because I could talk to anybody."

The issue arose because Pat Decker, Alice's boss at the time, decided that her car was not nice enough to be picking up such an important potential investor.

"Pat didn't think my little Honda Civic was an appropriate car," she recalled. "So, he gave me his Volvo sedan. So, I said okay. But it was a hot day . . . a very hot day. And because I had never driven this car, I couldn't figure out how to turn on the air conditioning. So, I was sweating to death by the time I got to the airport."

Luckily, or so Alice thought at the time, this was back at a time when you could park right near the terminal.

"So, I parked, and this really old man came out," she laughed. "At least he seemed old to me. And as we got to the car, I apologized for how hot it would be. But it got worse."

Alice had parked facing into a slot, which on face value was fine. The problem was, in this steaming hot car, she didn't know how to get the Volvo in reverse!

"I couldn't back it up!" said Alice. "I'd only ever driven standard transmission cars and this was a fancy automatic and you had to hit a certain button hidden on the gear shifter. We had to open the doors because I thought this guy, who's wearing a suit, was going to pass out or have a stroke! So, the two of us are sitting there and he's reading the Volvo manual while I'm trying to shift us into reverse, which we finally did. And though I got him to the meeting without killing him, I think that was my last time picking up an investor at the airport!"

Not long after that, Alice discovered her first true passion working with Mary Jane Conary in marketing.

"Working with Mary Jane was great," she recalled. "She was so driven. And what we were doing was really setting Kronos up for the future . . . getting the name out there . . . finding customers who were willing to serve as references. That was a position I enjoyed because I had a writing and editing background and it called for a lot of interaction internally and externally."

Honing her knowledge of Kronos and business in general empowered the former liberal arts student with essentially the equivalent of a business degree through on-the-job training.

"At some point we decided that we needed a marketing communications program," said Mark. "We hired a woman from the ad agency and she was horrible . . . I had to fire her. So, I turned to Alice."

"I liked marketing communications as I always enjoyed writing," said Alice. "But I didn't feel like I had enough experience to head it up. But I'd interacted with Mary Jane at the agency and I told Mark maybe she would be interested. And I was so happy when she came on board and became my boss because she was the nicest person and a great mentor. We definitely became a great team. She taught me a lot."

"She (Mary Jane) and Alice were an incredible team," said Mark. "They created testimonial materials and set up meetings with places like *Business Week* and *Fortune*. What they did together really helped Kronos take off."

By this point, as Alice noted, she had been in almost every potential position except coding. But she made a connection that, despite not filling out her resume with that job title, did bring her closer to that function.

"Yes, I met my husband, Jack Rich, at Kronos," Alice explained. "He was there when I arrived, working on coding with Larry Krakauer.

I got to know him initially when we were working on creating training manuals as a way to help customers solve some of their own problems with the clocks. Then I recall we really connected at one of the company's softball games."

"Yes," said Mark. "We used to hold these softball games against other local companies. It was a nice way to get outside, interact with other people, and have some easy physical activity."

"I believe Mark was the pitcher," Alice chimed in.

"I did pitch, but it was slow pitch," said Mark. "And truth be told I wasn't always there because I worked a lot of hours and most Saturdays and Sundays."

"You know," confessed Alice, "I think because we all worked so many hours and so closely that a lot of Kronos employees ended up dating each other at different times. Seemed like everyone went out with everyone, a little like the old TV show *Peyton Place*."

Others may have dated. But Alice and Jack married and are still together today.

"Another Kronos success story," Mark chuckled.

But beyond that, in first recognizing the potential in Mary Jane, Alice had shown another potential skill that she could apply to the company, one that would truly benefit from her gregarious nature.

"As we continued to grow, to the point where we needed to look beyond Boston for a bigger space, the idea came up that we needed what would eventually be called a human resources department," said Mark. "So, we looked around for someone with the right level of experience to do that function because I was getting too busy to always be involved in hiring, but there wasn't anyone out there that seemed like a really good fit."

The answer to this quandary had grown up in the same household as Mark, just not at the same time.

"The reason I had decided I wanted to set up an HR function had come about because Alice had sent me a memo proposing that it would be a good idea, because we'd grown to a level where we needed something more formal than me following my gut," said Mark. "So, after that early search I asked if she'd take it on."

Starting from scratch and with no formal training, Alice agreed, but only if she could do a couple of critical things.

"I did two things right away," she recalled. "I began taking classes in a certificate program in HR at Bentley University. And I joined the New England Human Resources Association, where I learned a lot and made some important connections with professionals who were willing to mentor me."

With the new function in mind, but still engaged in marketing because of the ongoing overlapping of functions, Alice also took every chance she could find while attending trade shows around the country to seek out HR pros for tips and techniques. Or she would interact with HR types from customers or even firms that were working with Kronos.

"I was just lucky that I found all these mentors that would give me lots of time and teach me how to handle employment situations," said Alice who cited one example, an employment partner at a well-known law firm that was handling legal affairs for Kronos.

"She was like this super smart Harvard law graduate and she took me under her wing," reported Alice. "And she taught me things like what to do with difficult employees."

Thus, Alice became empowered and proficient at her new position, so much so that she single-handedly managed the Kronos' transition from 30 employees to over 300 full-time staffers. And it wasn't just about filling empty desks, because the real challenge was in finding qualified people who would also be friendly, team-oriented, and able to hit the ground running, but would also contribute to the pleasant vibe that made Kronos a great place to work . . . and a great place to work hard.

"What I looked for," divulged Alice, "were people that knew how to share the shovel in the sandbox. They needed to play nicely and not hog sources or be egotistical. Because everyone that had started out there and stayed there for the ride, they were friends. We hung out together. Our social life was very intertwined with our work life. I mean, Mark used to have parties at his house and *everyone* came. But we also worked really, really long hours, including weekends, so you had to just really get along and be willing to play a part in that environment."

Alice herself decided her time at Kronos was complete a while back. But she feels quite good about her contribution to the company's success and, equally as important, the company culture she help perpetuate at a time when that could have gone awry.

"Kronos' success was everybody's success," she said. "But I think what makes me feel best are the times when I go back. And people seem to come out of the woodwork to give me a hug and say hello. They remind me that I hired them 20 or even 30 years ago. And they're still there!"

Among those long-time employees is Alice's husband Jack, still smiling as he walks to a meeting or to grab a coffee. But Alice reports that they seldom talk about work at home, preferring to focus on their own intertwined lives. But in this outside realm she found a way to take her own Kronos experience, and the experience she had as a youngest sibling to a great set of parents, and help others, as she now serves as a life coach and consultant.

"You know, as a coach and consultant I really work very intimately with people who are trying to resolve transition problems in their lives," she said. And I feel like I really have an impact. I think our parents set the foundation for that because they were always doing something in addition to the main part of living their lives, which was working, raising kids and taking care of family. They were always doing something generous and philanthropic. Helping other people was something that was instilled in all of us."

"Well, you really helped me," added Mark, "in many ways . . . really more ways than I can count."

"Well, you helped me, too," replied Alice. "You taught me about being bold and not necessarily doing things by the book . . . not being afraid to hire friends and relatives, which is not allowed at many companies. Many good people were hired who were related or friends. And all of those connections were good."

"And I helped teach you to drive," Mark reminded her.

"Yes, you did," she said with a laugh.

And, it should be pointed out, Alice is still a really good driver . . . one who now drives a car with an automatic transmission.

Chapter Seventeen

SWEET MELODY

Music is the universal language of mankind.

–Henry Wadsworth Longfellow

The universality of music is hard to debate. Fans in a country that doesn't speak English sing the lyrics to "Satisfaction" loud and proud along with Mick Jagger as if they were also from England. Many people first learn how to speak foreign languages by singing songs. Singers lose their accents when performing. And, in some remarkable cases, those with pronounced stuttering problems can croon as smooth as silk when given a melody and a beat to follow.

An oft-cited metaphor stems from the riff . . . as good news is considered music to one's ears.

Such a harmonious vibe must certainly have filled the Rich household back when young Jack Rich, a top-of-the-class student with big aspirations, opened his college acceptance letter from the Massachusetts Institute of Technology. Not only had Jack been accepted into the world's preeminent engineering university, he had also been granted a scholarship to attend!

Following a well-tread path, Jack opted into the Electrical Engineering program at the end of his freshman year, and it would seem that he was clearly on the road to future success.

Then came a summer of discord . . . a cacophony of conflicting thoughts. Though Jack's extraordinary intelligence ensured he was doing well enough in school, he finished his second year, packed his bags, and began making plans for whatever he might opt to do next. Like a needle scratching across a pristinely pressed 33-rpm record, Jack had decided to leave MIT.

"It's true," Jack would later confirm. "I had come to a point where I simply didn't care. Electrical engineering at that time was all about transistors. And I did not care how transistors worked. So, I was doing the work. But mentally I was dropping out."

Come that summer, he was dropping out in a way far beyond mentally.

"So," he continued, "I dropped out for the summer. But then, at the end of the summer, I didn't have a job as everyone else was going back to school. But I still had a scholarship. I decided to go back to MIT, but not to electrical engineering. So, I reregistered as a music major."

That might have resulted in a sour note. But the opposite was true. Jack had marched to the beat of his own drummer and away from something for which he'd decided he lacked true passion. He was close to alone, but not marooned solo on an isolated island, there being a near battalion of eight other music majors on campus at the time. And, for the record, his education, or at least one class within his undergraduate days, restarted the music.

Along with playing the piano and fiddle, singing in the chorus, and even performing at Boston Red Sox and Boston Celtic games, Jack took a computer programming class and realized that there just might be a job he'd enjoy . . . a job playing a different kind of keyboard.

"I'd minored in Computer Science," recalled Jack. "And even though I thought of myself as a musician, I needed a job. And I thought computer programming would be a good fit."

Though there was a dramatically long road ahead, advances in programming would eventually allow computers to become engrained in the day-to-day lives of virtually everyone on the planet . . . playing a universal language of their own.

Jack landed a job with Digital Equipment, and toiled there for about five years before switching to a smaller firm called Itek. After a short stretch, as several fellow employees decided to move on, Jack decided to engage a head hunter to find him a start-up where he could take his skill set as a diagnostic engineer. That head hunter sent him to Kronos, where the search was on for someone who could write solid code for its just-shipped time clocks.

He met with Mark and others and thought he could do the job. He just wasn't sure it was a good fit for him.

"Kronos, for me, didn't seem that promising," he said, "because they were only making time clocks. And I thought that would be boring. I mean, how much software could you put in a time clock?"

As it turned out, you could put quite a lot into a clock. And later, as technology changed the landscape time and again, there were ongoing and newer challenges that, for Jack, would span decades. As hardware begat software and systems integrated more and more solutions into the mix, the complexity of working within Kronos Engineering grew, and Jack grew with it.

"I didn't think I'd stay," he chuckled, "yet I've been at Kronos for 39 years now!"

"Jack made a very favorable impression," said Mark of that first meeting. "He definitely knew his stuff. And after finally getting our first product out the door, we knew we needed to improve the software if we were going to keep gaining in the market."

Mark didn't care that Jack had ultimately majored in music at his alma mater. As with many of his hires, he judged Jack's candidacy based on both his background and an intangible . . . whether he could both fit into the Kronos environment of the day and also bring the right skill sets to the party. There, Jack checked both boxes, especially on the programming side. True, Larry Krakauer had somewhat resolved early software issues, but even Larry knew that he needed someone more versed in software design.

"I joined right about the time they were shipping their first working product. I believe it was version 1.4," said Jack. "They'd had lots of previous versions . . . some that caught fire after cards got stuck in the puncher. The product was a strange big heavy box. And Kronos was very much a start-up atmosphere."

A start-up with carpeting that people tripped over, run by a man described by Jack's headhunter and others as a "hippie." But it was also a place populated by younger people filled with a start-up spirit, and, for Jack, a company that presented more and more complex issues, something he relished.

"The culture there was very young, with a real hard start-up focus. People worked hard and partied hard," chuckled Jack. "And, truth be told, the time clocks turned out to be very hard to program. So, it was not boring at all!"

A particular early challenge stands out as justification for both Jack and Mark.

"Jim Kizielewicz had been hired just after me," he recounted. "And he came to a meeting one week and told us that customers in health-care needed to be able to calculate weekly overtime but issue biweekly paychecks.

"Everyone in the room was saying it can't be done," Jack continued, "and I said, 'I can do it.'"

The new guy, the one with the musical background, would compose a solution. He would not become Bach, Beethoven, or even Charlie Daniels. Instead of composing symphonies or Top Ten hits,

Jack had found not just his niche, but a path he could be passionate about. He would become the Ada Lovelace of Kronos.

Computer historians will recall that the first computer to use punch cards as memory and a way to program a computer was created by English philosopher, mathematician, and mechanical engineer Charles Babbage in 1837.

Like many inventors of his era, Babbage had a hand in many inventions, each of which were created to solve a basic need. For instance, as trains were a main mode of transportation through farmlands in those 1800s, there were often delays brought on when cows and other animals would wander onto the tracks. So, to keep trains on schedule, Babbage invented what came to be known as the locomotive cow catcher, which was—and is today—the front part of a train engine capable of pushing things—even cows—off the tracks.

But Babbage's legacy innovation, much like later multi-inventor Thomas Edison's light bulb, was his Analytical Engine, the first computing machine to use punch cards as memory and a way to program the computer.

Through this innovative creation, Babbage would become known as "the father of the computer." But this designation would not have been possible without Lovelace, a mathematician who, by virtue of her work on Babbage's creation, would draft what is recognized as the very first algorithm created solely for the purpose of being processed by a machine. So, it was Lovelace who would become universally recognized as the "world's first computer programmer."

"I had come out of college thinking of myself as a musician who knew some programming. So, when I graduated, I was able to get a job," Jack said. "But, as time went on, I realized that, in programming, I'd found something I could be passionate about."

So rather than writing concertos or commercial ditties, Jack Rich, more than 150 years after Lovelace, would use his programming skills to be involved in most, if not all, of the myriad advances Kronos would take moving forward.

"I solved that healthcare problem," said Jack. "Then it was on to the next thing. I learned pretty quickly that software is never done. That was a mental adjustment I had to make . . . understanding that no one is ever truly happy because there's always something else to take you beyond what you just accomplished. You cross from one thing to another . . . building and building and building. That meant you could work and work and work. And that was the way it was early on with

Kronos. You'd come in on weekends. But you were never alone. But that's what you do if you have a really hard start-up focus."

From Mark's perspective, bringing Jack on board added a bona fide programmer to the mix at Kronos. True, Jack had only minored in Computer Science at MIT. He'd taken only four classes. But within those early stages of the evolution of technology in that period, programmers were realistically learning on the job, not in the classroom. So, Jack's five years at DEC, plus his other experience, and the intrinsic I-can-do-that attitude he displayed right away, made him valuable both right away and for the long haul.

"Basically," Mark would say, "Jack had a hand in every major advance we undertook. In fact, I think Jack is the only person to be involved in every technology advance made at Kronos!"

Initially, advancing meant putting more and more capabilities into the clock itself. At that point, the business plan called for clocks to be specialized according to a particular industry, like healthcare or manufacturing. And the focus was on the hourly workforce . . . called frontline labor.

The clocks were still much more expensive than their predecessors, but the cost savings associated both with monies saved by accurately paying employees the first time as well as in needing far fewer clerks to tabulate time was quickly proving to be a more-than-strong sales point, with a fairly quick return on investment coupled with increased employee satisfaction. But, as Mark made clear to the team, the best way to stay ahead was to continually innovate.

"We charged Jack to put more and more complex software into TimeKeeper," said Mark.

Jack felt that the next challenge would involve finding more ways to measure job costing . . . giving managers a view into how their workforce was performing on certain tasks so that efficiencies could be identified and implemented.

But Mark had other ideas. Kronos had demonstrated its value in healthcare and manufacturing. But there was another vast frontline kingdom to conquer.

"Mark was very fair," Jack laughed. "He said we promised you could do the job costing clock. But we also want you to do what's best for Kronos and work on the clock for retail. Put that way, what could I say to Mark? I got to work programming a clock for retail."

"That (retail) was the product that put us on the map," recalled Mark. "And Jack was a driving force behind it."

"Mark was right," Jack conceded. "We had a few clunkers. And job costing was one of them. We acquired a company that aligned with it. But it never sold. I always felt fortunate that I didn't do a job-costing clock. Retail was much more flexible."

Luckily, most of the sour notes were few and far between. And pursuing job costing went by the boards as the next technological wave came crashing ashore. Personal computers . . . PCs . . . had been around for a while with several niche players attempting to forge into a leadership role. But when powerhouse IBM came onto the scene in the mid-1980s, things changed quickly.

"No one took the PC seriously at first," reported Jack. "Then, when IBM moved in that direction, Mark said put the System 70 into a PC and let it talk to the time clocks. And the ability to network the clocks to the PC became Timekeeper Central, which quickly became our flagship product. And that was that. From that point forward I never had to write software for clocks anymore."

It wasn't a snap-of-the-fingers moment. There were still a lot of companies that had their own legacy systems. There were many different codes, customers using floppy discs. But in the PC, Mark and Kronos saw consolidation on the horizon and set the right course at the right time, again with Jack writing and rewriting an evolving code.

"There was always another thing, so we just kept on building and building and building," said Jack. "But the timing was right. People wanted open databases. But the PC was not easily accessible. So, client/server became the next big thing. Timekeeper Central took hold, but then along came this thing called the internet. And the internet was so big it had to succeed. So, we got on that wagon and that meant we had to rewrite everything in Java. So, Timekeeper Central and client/server became Workforce Central, which was our internet solution."

"Again," said Mark, "we had to evolve with the technology. If we hadn't done all that, we would have gone down the tubes."

"Fortunately," said Jack, "we always seemed to see the light in time. When Microsoft Windows came out, we had to have a Windows product. Luckily, client/server was Windows, so we could adapt. There's always a next generation thing."

For Jack, "next generation" is a term that holds a multitude of meanings. During those hectic-but-team-oriented early Kronos years, one major milestone was achieved unexpectedly. At a company softball game, his awkward and seemingly futile attempt at swatting away mosquitoes somehow attracted the attention of a coworker named Alice Ain.

"I think she looked up and saw this weirdo walking over to sit on the bench swatting at mosquitoes," laughed Jack. "But for some reason we hit it off."

They began to see each other socially. But even though their dating situation was not all that unusual within a fledgling company populated predominantly by younger people who worked long hours in close quarters—in fact, this was an easy recipe for the development of a few relationships—they tried, albeit unsuccessfully, to keep things quiet.

"There were people at Kronos who saw each other outside of work . . . dating," recounted Jack. "But we didn't want people to know. So, when we'd see each other coming or going through the hallways, we used to brush up against each other just a bit. Until one day we both moved over a little too far and walked smack into each other. Eyeglasses went flying. And I think, right about then we decided maybe it was too dangerous to keep it a secret any longer."

Mark had chosen to hire both Alice and Jack for the same reason, which was that each brought something to the equation that would make Kronos better. And he was right. Alice was a big part of early marketing efforts and later a huge part of starting the company's Human Resources function. And Jack, again, played a key role in software release after software release. Outside of work, they excelled through dating, wedded bliss, and parenting a son and twin daughters.

"I know some people thought that I did well because I was married to Alice," said Jack. "But the truth is that she did very well at what she did. And I did well at what I did. And we both contributed to Kronos' success."

Not everything went perfectly during Jack's multigenerational engagement at the company. One particular assignment stood out in his memory as a moment he was glad that he wasn't fired.

"I had been successful working in Major Accounts. But then I was coerced into heading Custom Accounts," he reported. "Truth be told, I didn't want the position because, to me, Custom was a losing proposition."

It wasn't that he'd been put in charge of a bad team. By that point in time, Kronos' engineering group had grown in size, stature, and domain knowledge.

"Software engineers historically had moved from company to company," said Jack. "And we had a lot of turnover at first. But we eventually wound up with people who stayed, and that was important because it's a pretty complex domain. So, domain knowledge was key."

But building custom software for a specific customer, even a giant of its industry with legions of employees, was simply not a profitable way of doing business.

"After a stretch the CFO looked at our numbers and I hadn't moved custom into the field," Jack recalled. "But I didn't get fired! I went back to a technical role, a role that I enjoyed, instead of managing. I like to think I survived being head of custom software."

Then there were the cool gigs.

"My best job," Jack remembered, "was as head of international. International at that time, where software was concerned, was so screwed up you couldn't help but do better!"

"Jack helped us a great deal in becoming a global company," lauded Mark. "In fact, Jack was invaluable in every tour he took on where the software was concerned."

As the internet begat the cloud, Jack again found himself in the mix. For that endeavor, Kronos assembled a group to work against . . . Kronos.

"The charge," Jack detailed, "was to invent a solution that would have the capability of putting Kronos out of business."

Code-named Falcon, the output of that undertaking would eventually be unveiled as Workforce Dimensions. And it has set the company on course to even bigger things, thanks to a music major from the world's preeminent engineering university.

"I never sought out to stay in one place for an extended period of time," said Jack. "Now I've been at Kronos well more than half my life. Why did I stay so long? Truth is I liked it, and then it became impractical to leave. I never found anything where I said 'That would be better' and so I stayed. Really, I'm not thinking about retiring. What would I do? I can't imagine not working . . . not when Kronos has flexible time. And the software . . . the software needs constant updating . . . "

Certainly, music to the ears of the engineering team at Kronos from the Music Man, a graduate of MIT who still has one incomplete on his transcript.

"Truth is," said Jack, "I never did finish that course on transistors. Lucky me . . . "

Chapter Eighteen

THE DISRUPTOR

Every success story is a tale of constant change. A company that stands still will soon be forgotten.

–Richard Branson

Sometimes, when change is required, there is an inherent need to blow things up. When those moments arise, it's important to ensure that collateral damage is minimized because, if everything is destroyed, you're starting from scratch. The trick is to hit the intended target.

Come 1989, Kronos was a fully formed, fully functioning, and increasingly profitable company. But the sales model that had created a groundswell of momentum was bogging down. As Kronos offerings became increasingly complex, it was hard for an outside sales force to keep up. Mark believed the solution was to transform a sales model that mainly utilized outside vendors to one based on building an inside sales force.

So, Mark was indeed about to blow things up. Naturally, then, he sought the advice of a bombardier.

Arnold Daniels had been one of those rare WWII navigator/bombardiers who was both skilled at his function and blessed to have survived through a harrowing thirty-five successful missions (with "success" defined in both targeted bomb drops and surviving to tell the tale).

Thirty-five missions might not sound like a lot to a lay person. But the average life expectancy of a bomber crew was fifteen missions. And gunners and bombardiers were specifically targeted by enemy craft, so their life spans were often measured in weeks.

Arnold and his crew were considered such a great team that the U.S. Army Air Corps (USAAC), having lost such a high percentage of lives and planes, wanted to better understand what made these particular bombers tick so that future crews could be assembled to replicate the squad's success rate.

To that end, Arnold's crew was put through a series of tests designed to determine both common and unique traits within the group, as certain characteristics would make for a good squad member, while others would separate the gunners from the bombardiers and the pilots from the crewmen.

What the USAAC was able to prove was that not all airmen were created equal. A series of questions pertaining mainly to math, vocabulary, and logic were asked against a ticking clock. The test was then scored and evaluated not just by numerical score, but by which specific answers were correct and which were wrong.

By establishing certain patterns, an individual's personality traits and decision-making ability could be identified. And those with certain abilities and potential could be assigned to areas where they stood the greatest chance of success. By ascertaining the right fit, at least metaphorically, round pegs were inserted into round holes while square ones were dropped into square-shaped receptacles.

Befitting his 360-degree view from below the belly of the plane, Arnold saw this behavioral assessment, Predictive Index (PI) exercise, from a broader angle. True, at the time these appraisals made for the formation of better and more successful combat teams. But, he reasoned, the same tool could be used on the ground, sans bullets and bombs, to build better business teams!

Not every postwar company bought into the concept. Those that did took some convincing. But Arnold persisted and by 1955 his version of the PI assessment was known throughout quite a few business circles. A few decades of success followed, as those organizations whose leaders opted in saw great value from their investment.

By 1989, with Arnold loosening his grip on his venture, he and entrepreneur Mark Ain happened to be neighbors. Mark explained to Arnold that he was positioning Kronos for an IPO. But he felt that to advance to this next vaunted level, Kronos needed to upgrade the hiring process for its expanding internal sales operation. Mark saw Arnold's now time-tested approach to employee evaluation as a way for his company to identify the right people to comprise this expanding sales force.

Arnold was intrigued. He had quickly developed a fondness for Mark and the way he had fashioned a veritable "little engine that could" in his fledgling Kronos, so he agreed to advise from the periphery. The ex-bombardier felt Kronos would be in better hands working directly with his protégé, Stan Kulfan. Arnold would be available to serve as Stan's sounding board and guiding force, while

Stan would be empowered to handle all assessments and make all recommendations.

And thus, like many of the seemingly unlikely pairings within Kronos, began a long-lasting and extremely fruitful relationship.

"You helped change the company," Mark would later tell Stan.

"I was certainly a catalyst," recalled Stan, "and eventually I guess you would call me a trusted partner."

Stan's first assessment came in advance of administering any test.

"Mark was an interesting case unto himself," said Stan. "He was kind of a hippie guy on one hand. He was also very into computers. I knew Arnold had developed a quick interest in both Mark and Kronos. And Arnold was the type of person who quickly and accurately judged people. And more than anything, I could tell right away that Mark was one of those unique entrepreneurs who was willing, and even excited, to get a different perspective."

"What Arnold had developed, and what Stan brought to us," explained Mark, "was a tool that enabled us to understand each other . . . something that showed us how to best relate to each other. And with that understanding, we were able to think about, as a team, how to get to the next level."

Like many a pathway to a solution for a complex problem, the catalyst was something that was seemingly innocuous.

"Getting to the Predictive Index is deceivingly simple," Stan divulged. "It all starts with having each participant answer a two-page questionnaire."

Two pages! Questions varying from math to history to science . . . fact-based and intuitive . . . all adding up, not so simply, to much more than a numerical grade based on correct versus incorrect answers.

"The ultimate test is always in the data," explained Stan, "not how many questions you answered correctly, but *which* answers were correct and which ones were not. The real answers were in the data."

Stan's approach was not as originally set out. Yes, he would assist in coming up with a way to assist in the creation of an internal sales force. His focus would be on developing a methodology designed to find people with the right mix of sales hutzpah and technical knowledge. But first he both wanted to prove the mettle of his method and get a sense for the management team with which he was about to work.

"I said to Mark, let me show you how this works. So, the first workshop I set up was with the management team," he recounted.

"Arnold always believed that the best way to prove that the PI was an accurate tool was to take a group of people that the CEO or founder of a given company knew well and assess them," said Stan. "He believed that if you could prove the insight gained you could more easily understand the power of the Predictive Index and things would take off from there."

So, without knowing any of the other members of the management team personally or professionally, Stan set about administering and scoring the tests of then-current key Kronos leaders off site in a hotel training room. What Stan gleaned from those tests and meetings was a game changer that would not simply affect how Kronos would hire a superior sales team. Rather, many roles within the leadership team were about to change . . . including Mark's . . .

"As we got into it," said Stan, "lightbulbs were flashing and everyone was really getting into it. And as everyone started to see the results, the skepticism disappeared."

"It was amazing," said Mark. "Stan came in and it seemed like every evaluation was right on target. But the real revelation was that I'd brought him in to show us how to fix sales. But he quickly was able to show how this wasn't just about sales."

"There were several things going on that were holding Kronos back," said Stan. "But it started at the top. In order for Kronos to change, Mark had to be willing to change!"

Luckily, true to his initial read on Mark, Stan quickly found that he was partnered with the rare chief executive who was more about the company than himself.

"I can't tell you how many times the lead guy doesn't get it!" Stan lamented. "The vast majority of leaders, tens of thousands over the last thirty or forty years, they all started the same way. Mark built something. But it wasn't just Mark, and he knew that. But even though his natural instinct was to be 'the guy,' he understood he had to leverage people to get to that next level."

In a sense, that starting point was about Stan detailing to Mark how his early instincts had gotten him to this enviable juncture, but that, in order to keep moving past a very definitive line in the sand that most fledgling companies fail to cross, it was necessary to now fight those same instincts.

"His instincts . . . the team of people he hired without any Predictive Index . . . his instincts were great," Stan said of Mark. "But, by the time I came along, Mark had moved into the role of problem solver. He needed to delegate that role to other people because, as the PI showed, Mark was a disrupter."

The diagnosis made sense. Mark had, after all, founded Kronos on the very concept of disrupting an industry. The success Kronos had enjoyed to that point was predicated on that premise. But the point where Stan had gotten involved was a critical crossroad, one where many a potential longer-term company ran out of fuel.

"This was a classic case," said Stan. "This was what would happen when companies reached around $20 to $25 million in revenue . . . when they hit 100 to 150 employees. And all of a sudden what had been working doesn't scale. It's a tough point, because to that point things have been going well."

As Mark had always said about the solutions, "If it isn't broke, fix it anyway!"

"Stan was right," Mark confessed. "This wasn't about changing how we sold the product. It was about how we ran the company."

Truth was, despite all the esprit de corps that successful start-ups and young companies can enjoy, and perhaps even less visible at Kronos because of the high-caliber characters Mark and the team had brought on board, there were cracks in the Kronos fuselage that would need to be fixed in order to travel ever higher.

Internally, there was a degree of friction between respective departments. The board, which mainly consisted of investors at that juncture, had strongly suggested that Mark bring on a chief operating officer, thus marking the first major hire that had been dictated to him. This individual's style was not necessarily wrong. But given his background, a clear focus lay in favor of manufacturing versus sales. So, the edict was that sales should follow manufacturing.

Problem there was that the nature of Kronos sales . . . specifically because the company was a provider of solutions to other companies . . . came at the end of fiscal quarters, and mainly at year end.

Within that backdrop, a problem had been arising within order processing, where a distrust of the sales arm began to fester.

Rather than confront anyone directly, including the COO, department heads took to cornering Mark to curry favor for their own group. This meant Mark was constantly putting out fires, smoothing egos, and establishing and maintaining the peace, all of which collectively left little, if no time, for disrupting an industry.

To prosper into a next level company . . . one with legs for the much longer haul . . . would mean change . . . for everyone from top to bottom.

"Mark really did use me to proactively change the culture at Kronos," said Stan. "But it had to change! That meant frank conversations with everyone. A couple of folks left, the COO among them. But we needed to scale the decision-making process. Mark needed people he trusted. And, it was soon realized that the company needed a president."

By president, what Stan ultimately meant was that Kronos needed a role most closely associated via the predictive index as a "bridger" or a strong mediator. Within the executive ranks, Pat Decker and CFO Paul Lacy would both step into roles designed to lead in that regard. And this seeming dilution of power created the exact prescribed role for Mark the Disruptor.

"With this change, the organization was set to climb higher because of everyone's more clearly defined and well-matched roles," said Stan, "and because Mark was now free to be Mark. He was no longer a referee. And, for Kronos to be successful, you didn't want Mark to be bogged down in the day-to-day details. You wanted him free to shake things up. Then, by maximizing Mark in his role, the road was clear to maximize others as well."

"It was amazing how quickly the company flipped," said Mark. "But that was just the beginning."

This was the ultimate truism. Because as Kronos was changing as an organization, the universe in which it was thriving was again undergoing volcanic change.

"Kronos was, or at least had been, a hardware company. We made clocks," said Mark. "So manufacturing was key. But it was becoming clear very quickly that software was becoming more and more important. So, we needed to lead in that arena so that we didn't one day soon find ourselves behind some other upstart company."

"The basic idea was to make Kronos a sales-driven, high-performance hardware and software company," reported Stan. "And here's the trick . . . once you fix sales and manufacturing, then every group within the team has to see it that way. Everyone's bottom line had to be sales. Everyone had to be a closer!"

That meant aligning sales and order processing. But then the service organization was slipping behind, so they had to step it up.

"Once you make that turn to being sales-driven," said Stan, "it has to roll throughout the organization. There were no excuses. You couldn't blame someone else. It was everyone's job! Mark was aggressive. But when push came to shove, it was all hands on deck!"

Revenue goals were met or exceeded. The IPO objective was met and Mark and his parents got to ring the opening bell for the Nasdaq. Kronos flew higher when, after missing an early chance (hitching the wagon to DOS instead of Windows at one early juncture), the software ride started to kick in. And the reshaped organization would go on a run of 120 consecutive quarters/year over year with profitability, a run exceeded in that time frame only by a little outfit called Microsoft.

Through it all, Stan maintained his independence . . . the rare major contributor who was never an official employee of Kronos.

"That's true," Stan stated. "But I think I attended every executive board meeting at Kronos for over 10 years."

"Your major contribution was teaching me and the executive committee members how to change our style in order to grow into a much bigger company," said Mark. "You changed the way our senior management team performed by showing us how to run the company better."

"Leaders need to understand themselves, and each other," surmised Stan. "You've got to understand people to leverage their talents. Then you have to get them to do the same thing and leverage their people. The only way you get to make the correct decisions together is to understand where everyone is coming from, know the differences in perspective, and then realize that success is about everyone aiming at the same goal. Mark and the people at Kronos understood that."

Together, Mark and Stan realized their respective goals. At the time of their first meeting, Mark wanted to pilot a company that he could take public in short order. But his goal was steadfastly set on a long-term horizon. Stan took great pride in serving as a catalyst for change. He had the tools in hand, but success was a blend of skill and guts.

Or, as Arnold would have likely stated, it was a matter of carefully getting the plane out over the intended target, opening the bay doors, and then bearing witness to a direct hit before turning back to refuel for what would undoubtedly be yet another arduous, and great, adventure.

Chapter Nineteen

ACCENTUATE THE POSITIVE

We shall never know all the good a simple smile can do.

—Mother Teresa

A lot of interesting partnerships began in 1994. Samuel L. Jackson teamed up with John Travolta in *Pulp Fiction*. Michael Jackson wed Elvis Presley's daughter Lisa Marie. Jeff Gillooly, ex-husband of U.S. Olympian figure skater Tonya Harding, teamed up with bodyguard Shawn Eckardt to disable Harding's chief competitor, Nancy Kerrigan.

None of those couplings yielded the desired results.

But, in a coupling that would grow and prosper and stand the test of time, Cheryl Ferruccio, leading with a smile, interviewed for, and was subsequently hired, as an executive assistant at Kronos.

Mark Ain had an admin at the time. So, Cheryl's first assignment was assisting CFO Paul Lacy. But given both how much time Mark spent in Paul's office, and Kronos' relatively small size at the time, Cheryl often had crossover duties that had her also working closely with the company's founder and CEO.

"Initially I worked for Paul," detailed Cheryl. "But this was just shortly after Kronos had gone public, toward the end of 1992, so Mark and Paul, and also Mary Jane Conary, were always preparing to go on the road on trips to meet investors. So, I worked very closely with them. Plus, especially in those days, when Kronos had only about 400 employees, everybody absolutely knew everyone in the company."

More than simply someone who took notes, Cheryl quickly became a cog in the core investor-relations team, compiling key data, and even had the responsibility, and honor, of calling in the quarterly numbers to the folks at Nasdaq.

"That was the beginning of a stretch where Kronos went up in earnings and profitability quarter over quarter and year over year at a pace that separated the company from most other companies," said Mark. "In fact, after a while, the only software company with a longer streak of consecutive quarters of revenue and profitability growth was

Microsoft. And Cheryl was a key player in terms of getting us ready to go out and leverage those numbers to attract more investment in the company."

"I loved those days," Cheryl smiled. "Kronos was successful. But it was also like family. We were putting together road shows and acquisitions. We all celebrated those successes, but we were also constantly reinventing ourselves. Mark always had a vision to advance and adapt. We kept up with the times, but always looked ahead. Mark was never happy maintaining the status quo."

After about a year and a half with the company, Cheryl's path ahead changed for the long haul. Mark's admin stepped down. And as he often did, Mark looked inside for just the right person to fill that important role. But he didn't need to look far. Cheryl was already just around the corner from the actual corner office.

"It was a great opportunity and I never looked back," she said. "When I started it was just time clocks. Then the shift was to software . . . then Windows . . . then the web. Sometimes it was hard because you had to keep up with constant change. But there was a vibe. Revenue always grew year after year after year."

"What I enjoyed about Cheryl was that she was always so positive," recalled Mark. "And she didn't hold back. She always told me what she thought. That was a part of our culture. You were encouraged to say something. Don't hold back."

Of note, Cheryl assisted within executive communications. But she was also the voice of Kronos HQ.

"That's true," laughed Cheryl, "but Mark was very approachable. The whole executive team was approachable."

Over the course of 25-plus years, Cheryl would become a confidant, scheduler, assistant, sounding board, and friend—the jelly to Mark's peanut butter . . . the eggs to his ham—as she's played a key role and witnessed the company's steady climb from mid-level to multi-billion-dollar industry leader. Along the way, her interaction with Mark and her fellow Kronites has given her a unique perspective on both founder and company.

"Looking back, one thing that stands out now," said Cheryl in retrospect, "was that when we were a smaller company, nobody knew who Kronos was. I remember memorizing our elevator speech that explained what it was we did."

At the time, Kronos was essentially a "company's company." Its consumers were not everyday people. You didn't go to the local department store to buy a Kronos product. Yet, in the backroom of that department store, or hotel, or manufacturing facility, the employees who worked for that company and were signing on and off from work were likely doing so through a Kronos system.

"It was always funny when you'd see a picture, or a scene in a television show or movie, and you'd spot a Kronos clock in the background," recalled Cheryl. "I'd point right away and think *There we are!* And people were interested when I would tell them what Kronos did. Though, of course, as the company grew and advanced into providing more and more services, people started to understand who and what the company was about."

Like calling Nasdaq with the revenue and profitability numbers, Cheryl also enjoyed her interactions with Kronos customer at events like the company's user conference, an annual gathering of customers, many of whom interacted with the systems on a daily basis and attended to stay up to speed on product advances and pipelines.

"I always enjoyed interacting with customers," she proclaimed. "And the user conference, what's called UKGWorks today, was my time to be in front of the customers. I loved meeting with and talking with those people because they were always so excited to talk about how Kronos was helping them better do their jobs by better managing their people."

As with any journey of an extended duration, there were a couple of bumps along the way. For Cheryl, the ones that stood out were the first time the company did not report earnings growth after the economic downturn of 2008, the immediate stretch after Kronos transitioned from a publicly traded company to a privately held organization, and most recently the merger with Ultimate Software, a move that almost literally doubled the size of the company overnight.

"There were a couple of times that were scary in that I didn't want to lose my job at Kronos," she said. "I always believed I could find another job. But I'm not sure there are a lot of places out there like Kronos. Here there has always been the same basic direction . . . customers first, employees first, families first. So, even as the company grew and we had more and more people on board, everyone still supported that same mentality. And I think Kronos will always be that same company because everyone here really supports each other. I think it would be hard to get a job with a company that had a similar culture. So, I wanted to stay here. That's for sure!"

The other major transition for Cheryl came about when Mark decided it was time to step down as CEO.

"Mark stayed involved and quite busy after he stepped down as CEO," Cheryl said. "He was executive chairman of the board. And he was still traveling. So, he stayed busy and involved, just in a different way."

And now that Mark has relinquished more Kronos-related responsibility of late . . .

"I think a lot of people don't know how busy Mark is in retirement," remarked Cheryl. "Mark very readily shares his expertise with other boards and other companies. He's very involved in helping out young entrepreneurs and others, especially through the University of Rochester and other organizations.

"I really wish more people knew how much good he's doing," Cheryl continued. "In fact, I'd say it's amazing how busy he stays, and how much he shares. I think when people heard he was relinquishing more of the day-to-day that he was just going to sit still. But that's not him. He is, and has always been a great boss, and a better person. And I'm extremely lucky to have worked for him all this time. Extremely lucky."

Chapter Twenty

BACK TO THE PRESENT

The only place where success comes before work is in the dictionary.

—Albert Einstein

Recently, Mark was at one of those big box stores (yes, a guy with his net worth still buys in bulk!). After paying, a "receipt checker" was making sure he'd paid for all his items when Mark asked him how his experience was when interacting with the company's time and attendance system.

Taken aback a tad, the man said it was very efficient. In fact, he was impressed with the system's capabilities, which included the ability to request time off and check on accrued vacation days. Seeing his quizzical look, Mark explained that he was the founder of Kronos, the company behind that system.

Before he would let Mark leave, the man asked for permission and took a selfie with Mark, explaining to him that not only would he be sharing the picture . . . he intended to print a copy to keep beside the store's Kronos time clock.

Both men smiled for the tiny lens within the employee's cell phone, and Mark's face shone with a cool, perhaps ex-hippie-ish glee. After all, he'd spent a decade building a company, and then decades running it, in relative obscurity in comparison to many of the other technopreneurs of his generation. The fact never bothered him. Success was the goal, not fame. Kronos did the dirty work of its customers, now numbered in the tens of thousands and located just about everywhere on the planet where people are hired, scheduled, managed, and paid as employees.

The company he'd founded had changed the business paradigm in a dramatic way. Kronos solutions enabled its customers to focus on their core competencies because Kronos solutions streamlined the day-to-day running of these businesses.

The run-up to global status had been, if not meteoric, steady and consistent, especially when compared to most other technology-driven organizations of the day.

Under a game plan designed for a marathon versus a sprint, Mark focused on continuous improvement and expansion of the Kronos solution set. True, it took over 10 years and many millions of investor dollars before a profit was finally etched into the books in 1988. But once the butterfly had emerged from the cocoon, Kronos stayed airborne.

In the build-up to profitability and beyond, the company showed year-over-year revenue growth, quarter over quarter, for 28 years, a record surpassed only, as previously noted, by Microsoft. Yet unlike Microsoft founder and Executive Chairman of the Board Bill Gates, Mark Ain retains, and relishes, near-complete anonymity whether the founder and Executive Chairman of the Board of Kronos is filling a shopping cart with wholesale-priced household items or picking up a pastrami and rye from his favorite local sandwich shop, where he has to tell the cashier his name is Mark in order to receive his order.

For Mark, the level of accomplishment that came from focus and hard work had more to do with the team than any individual, so his greatest sense of accomplishment was in fleshing out his management roster.

"Truth be told," he said, "I took chances on people other companies wouldn't take chances on. A lot of other companies had a very strict profile for whom they would hire. I believed in people, and in my ability to sense when they would be a right fit. On paper maybe some of them weren't by-the-book hires. But a lot of them worked out very well."

A testament to the success of his somewhat unorthodox approach to constructing what would become the Kronos management team can be found in the revenue and profitability numbers, but the underlying foundation was bolstered by the industry-bucking trend of longevity within that core group.

"Once we had the core players in place, we retooled the way we operated as a management team," said Mark. "Responsibilities changed, including mine, and we settled in and really began to take off. We met regularly and reset our quarterly and annual goals each time. We got better and better at executing the game plan as we kept redoing our products. As technology changed, we changed with it."

There were missteps along the way. And not everyone stayed in for the long haul. But as the clocks became more sophisticated with each iteration—breaking down and providing a way to control labor costs, making payroll more efficient and less costly—customers became easier to acquire.

With the dawning of the PC era, Kronos was able to empower companies to both eliminate paper trails and centralize operations. And with the dawn of a true world wide web, Kronos separated itself from the vast number of dot-com posers by actually finding ways to incorporate the internet as the backbone of its most recent and burgeoning suite of solutions, designed to broaden the company's reach beyond Time and Attendance and into HR and Payroll.

When the company went public in June of 1992, the swelling number of customers included some of the biggest companies in the world.

As the original offering at $12 doubled to $24 a share, Kronos was shipping over 25,000 units, yet catering to only about 15 percent of potential customers in the U.S. market alone. A vast universe of opportunities lay in sight. Kronos' eyes turned toward Canada, the UK, Australia, and beyond.

"With the right people in place, we rolled through the '90s," reported Mark. "We got to enjoy the success because we didn't have to focus our energy on success. It wasn't easy. It was still hard work. But we didn't need to worry about making it. We were there! Success became part of the routine. Even so, once we finished one product or initiative, it was on to the next one. Pressing ahead was also a part of the routine. The mantra was to 'Never stand still.'"

With product development achieving milestone after milestone, the business focus widened to include the expansion and maintenance of a world-class service organization. Implementation and training were critical at both ends of the equation, as new customers wanted systems up and running as quickly as possible so they could realize a return on their investment in Kronos. And, as more and more organizations were able to report both rapid adoption and return on investment, many within six months, utilizing referenceable customers was becoming an increasingly indispensable aspect of the sales process.

The trajectory was clearly aimed upward, with an updraft under the wings for good measure. Mark stood on the bridge believing in the crew he had aboard.

"We had a great management team," said Mark. "We had a wonderful Board of Directors. And as we'd expanded as a company, we'd adhered to a basic principle . . . we were looking for people who knew how to roll up their sleeves. We had challenges like any growing organization. We had some turnover . . . some voluntary . . . some not-so-voluntary. But for the most part, together, we kept moving in the right direction."

Soon, Kronos was earning awards for outstanding service even as its reputation for being a great place to work continued to grow as well. This highly acknowledged internal employee focus contributed, in turn, to a much-higher-than-normal record of employee longevity, something that brought even more fuel to the Kronos engine.

"When we first started," Mark explained, "timekeeping was a pretty straightforward concept. But as the technology improved, and we improved the platform with it, our solutions had become much more complex. Having so many employees with deep domain knowledge was very advantageous because they were up to speed."

Not just up to speed, but accelerating into other facets of what would come to be known as Human Capital Management. By adding and constantly improving solutions that covered functions like Payroll, Scheduling, Absence Management, Labor Activities, Benefits Administration, and Talent Acquisition, Kronos was able to dramatically lessen the burden of managing employees so its customers could reduce costs associated with employees, ensure compliance with state and local regulations, and focus on the core mission within their respective business models.

Likewise, the Sales organization grew from different sources and in multiple ways. The organic recruiting and hiring of individual salespeople coincided with expansion of other departments, while Kronos also acquired and absorbed many of the employees of former Kronos third-party vendors from around the country. These employees also came to the plate with at least some previous industry experience, thus shortening their learning curves. In step with many of the previous members of the Kronos family, many of these "newbies" would opt to stay with the company for decades versus years.

"I personally had a hand in those dealer acquisitions," recalled Mark. "And we kept quite a few people who really worked out. A lot of them were good people. And the third-party relationships had proved fruitful. But by that time, with the advancing technology, the objectives for Kronos and our dealers no longer matched up. We had a model that we knew worked in terms of training and support, so bringing the sales force in house was a very logical and necessary step."

The other logical step lay outside the United States. With the internet bringing the world closer together on a virtual stage, with trade treaties being set up that all but ensured that a global business world was on the horizon, and because Kronos was doing business with more and more international companies, it was time to look beyond the borders.

Kronos opened its European-based office in the United Kingdom, and followed suit not long thereafter half the world away in Australia.

"We opened an office in the UK, and everyone said we could never do a good job in Europe," said Mark. "But that office thrived and so did our office in Sydney."

The transition from the high-flying era of the 1990s and into the new Millennium found Kronos entrenching as a global player with offices in more European cities, Mexico, Canada, and Australia, with an eye toward India and China.

Entering that decade, his fourth at the helm of Kronos, Mark realized that a transition at the top of the house would likewise be beneficial.

This global expansion was not simply about bringing solutions to other countries. Rather, as Kronos opened offices around the world, the focus on hiring the right people and then ensuring those new Kronites felt valued and respected was a key factor in growing the company. And Mark took great pride in the fact that in virtually every country where Kronos had a significant office presence, the company was recognized as a place where employees were valued. This translated to honors in multiple years in the United States, Australia, Canada, Great Britain, India, and Mexico, with specific categories like Best Workplace for Women, Best Workplace for Mental Wellness, and Best Place to Work in Social Responsibility and Quality of Life for Its People.

Success on all these levels, as it seemed over and over again for Kronos, meant the time was ripe for change. But rather than expansion of its suite of solutions or geographical reach within a global marketplace, this change was internal.

Pat Decker was retiring from his role as President and COO. Paul Lacy would continue as CFO and assume the role of Executive Vice President. Mark then continued serving as Kronos CEO, while Aron Ain would step up and into the role of Chief Operating Officer.

"Pat decided to retire," said Mark, "and my brother Aron was his obvious successor as chief operating officer. This structure worked extremely well for a stretch. But even then, to be honest, I really was very, very tired of working six days a week."

Tired, but surrounded and bolstered by a solid leadership team, this structure performed well, and Kronos continued to prosper and grow until Mark arrived at a not necessarily unexpected inflection point.

"I'd talked about stepping down before," said Mark. "But toward the middle of 2005, I decided this was the time. I was 62 years old and, frankly, tired of the pressure. It was time for a change."

The year 2005 then became another of Kronos' corporate milestones. The company Mark had dreamed of somehow guiding to $100 million in annual revenue reported $500 million in revenue that year, while also notching its 100th consecutive quarter of revenue growth. And on October 31, exactly 28 years after officially founding Kronos, Mark Ain, the only top dog the company had ever known, handed the role of chief executive officer over to his youngest brother Aron.

Mark would assume the role of executive chairman of the Board. But with his type A, hands-in-every-pot mindset, it was hard, at least initially, to not intercede in the day to day.

"I quickly learned that Aron's management style was as different from mine as night is from day," he laughed. "But the reins were his and I had to adjust to that and make a big change in my style."

A contrast in style perhaps, but the mindset of constant innovation and moving forward was still the charge of the day regardless of which Ain was at the helm. To ward off niche competitors, Kronos realigned itself into various verticals, essentially developing deeper industry-specific expertise.

In 2006, the first office was opened in China. Then 2007 saw the first office in India and, in a dramatic plot twist, after a highly profitable fifteen-year run as a publicly traded company, Kronos was bought out by a private equity firm that, quite understandably, structured the deal in a way that kept most key management members in place. The stock market collapse of 2008 gave brief pause to the consecutive quarter of revenue streak. But having gone private just prior to that meltdown left Kronos strong.

"Aron proved to be the right choice," Mark confided. "He not only kept things moving . . . he took the company to yet another level. The world was scared at that point. And we missed that one quarter. But we were still positioned for success."

With the focus still on innovation, Kronos invested in mobile, cloud-based solutions. In 2014, annual revenue surpassed $1 billion, while in 2015, the one-millionth time clock shipped. As 2016 dawned, Kronos hired its 5,000th employee. And with customers flocking to the company's latest cloud solution, the future stays bright.

Again, in transition, Mark stepped down as executive chairman of the Board as Aron stepped into that role as well.

"I'm minimally involved now. But since I don't have anyone to answer to, businesswise, I'm more effective now that I've stepped away," he said.

More effective, because while he stepped away from most of his official titles, he still retained his unofficial and lifelong role as a disruptor of an industry, and of the company he created . . . against all odds.

Chapter Twenty-One

WHAT WOULD YOU DO?

Before you are a leader, success is all about growing yourself. When
you become a leader, success is all about growing others.

—Jack Welch

Mark had encountered many obstacles during his professional tenure. But the ascent of Kronos, under his guidance, had indeed propelled him to a summit of his own definition. The company became known as much for its welcoming culture as for its commitment to innovation and customer satisfaction. Milestone after milestone had been achieved even as the proverbial bar was set and reset higher and higher each year. And Mark had finally became comfortable with realizing that easing away from the daily grind was a destination now loaded into his mental GPS.

But what would he do? What would he do now that he'd achieved a lifelong dream of founding his own company and subsequently guiding it (with help from many teammates) to near-unprecedented levels of success?

Mark's answer was, as was often the case, multitiered. After so many years of six- and seven-day work weeks, he was clearly ready to relax a bit. And he wanted to enjoy his family more. But, amid all that, another spark was heating up inside his gut. There had to be a way he could give back.

Undoubtedly, he believed, there had to be others within a next generation who might consider following a similar entrepreneurial path to his for whom he could provide counsel and inspiration.

Since the dawn of the concept of work, there have been entrepreneurs. From bygone eras, Ben Franklin, Thomas Edison, and Henry Ford come to mind. The names Steve Jobs and Bill Gates would populate more modern lists. But for every idea that came to fruition, there were untold numbers of concepts that died inside someone's head. Most often, it wasn't that an idea lacked merit. Rather, many would-be success stories were doomed due to a lack of direction and support.

Mark's level of success had ensured that he could opt to simply kick back. He could improve his tennis and squash game, attend sporting events in really good seats, or take long walks on the beach with his wife Carolyn. And he committed to doing those things. But he also decided that he would do more. He would find a way to promote the next generation of game changers.

Mark thought back to MIT, a burgeoning bastion of independent thinkers. But though he'd benefited greatly from the friendships he'd made during his time in Cambridge, Mass., and though MIT had reached out to him on numerous occasions during Kronos' ascension toward success, he believed that it was ultimately his MBA experience at the University of Rochester that had provided him with the tools and the opportunities to develop into a business leader capable of forging his own path as an entrepreneur.

"The University of Rochester granted me opportunities and direction that proved invaluable to every aspect of my professional career," Mark said emphatically. "All my classes were excellent. Each of my professors truly taught me about the world of business. One professor, who taught Organizational Development, regularly invited all his students to dinner with both him and many of his successful friends . . . executives with Fortune 50 companies. And you could learn as much at one of those dinners as you could in a semester of classes!"

In 2005, Mark had been extended a massive honor from the University of Rochester. As part of its fortieth anniversary celebration, the dean of Rochester's Simon Business School asked the alumnus of the program's first MBA graduation class to be the commencement speaker.

Mark delivered a speech that not only earned him accolades from all in attendance, but also provided inspiration for the speaker himself. And, after his talk covered what he saw as his own motivations for bucking the tide of the time and striking out on his own, Mark took a closer look at the programs offered at Simon and that spark of interest quickly became a fire.

"They (The Simon School) had an entrepreneurship program," recalled Mark, "but it was run by people whose experience was in large companies. In reality, they had little or no experience in actually being entrepreneurs! How could you run a successful program?"

The University of Rochester had actually started teaching a class in entrepreneurship back in 1978. But the path had been rocky. The initial professor did indeed have start-up experience. But neither he nor his philosophies meshed with those of the other professors. And after a few years of acrimonious existence, that teacher was gone and the program essentially went dormant.

Sometime later, the right person in the form of Duncan Moore came on board. But Duncan immediately saw issues coming from opposing sides of the university.

"In 1988," Duncan recalled, "I started a course called technical entrepreneurship. The class was drawing students from both the school of business and from the engineering school at about a 50/50 ratio. But this was absolutely hated by the faculty of the business schools, who didn't think it was a legitimate business class. And maybe not so surprisingly, the engineering school didn't think it was a legitimate engineering class. So, in its own way, it was a very controversial course at the time."

The disconnect made filling the class difficult. Duncan persevered for a couple of years, but then took the reins as dean of engineering, before eventually leaving Rochester for a stint in DC serving as a science advisor to the White House, where he became in charge of a committee tasked with fixing the then-hobbled Hubble Telescope.

Upon Duncan's return to Rochester in 2001, he discovered a paradigm change in the way the school viewed entrepreneurship.

"When I came back from the White House in 2001, I proposed starting the course again," said Duncan. "But this time the deans thought this was the best thing since sliced bread."

That point in time came about in the wake of many companies being created in the dot-com era and other mavericks through the 1990s, as well as the longer-term success of early tech companies, among them Apple and Mark Ain's Kronos. The tech field in particular had seen a groundswell of start-ups, with the term technopreneur slowly finding its way into the vernacular of many institutions of higher learning.

Duncan was teaching his class. But, when a grant empowered and emboldened the university to actually create a program dedicated to entrepreneurship, he was, ironically, not the initial choice.

Moving back to 2005, there was Mark. Rochester was asking him what he wanted to do to make the university better equipped to educate its soon-to-be new leaders of the twenty-first century.

"I said I wanted to be involved with the entrepreneurship program," said Mark. "But I didn't just want to make a donation—I wanted to make a difference."

Mark once again embraced the role of disruptor. He was clear to the powers-that-be at Rochester that, after some evaluation, he believed the university's Entrepreneurship program as offered was simply not doing enough.

As an initial push, Mark provided funding for a business model competition which has borne his name since 2005. He then began the process of fully evaluating the existing program with, as he had with Kronos, an eye toward continuous improvement. Again, he wanted to make sure that growth was for the right reasons. It wasn't going to be about making one whopping donation. It would be about building momentum and taking the right steps at the right time.

"To be honest, I wasn't impressed with the way it was being run," he recounted. "At that time, Rochester was definitely not leaders in this field . . . not at all! But that was before Duncan took over."

The individual who was initially charged with running the program departed Rochester in 2006. And after a brief search, and a little prodding from Mark, Duncan was given the opportunity to take the reins. After a couple of early meetings that included Mark, Duncan knew the chance to take everything to greater heights was there. And after meeting with Duncan, Mark knew he had a champion he could back. Mark began spearheading a long-term capital campaign.

"Mark getting involved moved us to an entirely different level," said Duncan, who would eventually be named Vice Provost for Entrepreneurship, a lofty title for a program that all but totally lacked support in its initial phase.

"I look at entrepreneurship at the University of Rochester in three stages," Duncan detailed. "From 1978 to around 2000 the entrepreneurship program remained largely dormant, and things did not go so well. Then, we received a grant from the Kauffman Foundation in 2004. That grant period that ran from about 2004 through the adoption of the strategic expansion plan that Mark drove, where we garnered financial support and began to build success upon success. And then the period from 2014 on, where the program became not only a success, but also nationally and internationally recognized . . . that began a golden period that, in many ways thanks again to Mark, continues to this day.

Mark's efforts resulted not only in the initial programs and classes being restarted, it also led to funding for things like the creation of student incubator space. Soon, students were entering the Mark Ain Business Plan Competition and realizing the value not only of the monetary prize, but also of exposure to Mark because, while it was true that Mark made substantial monetary gifts, he also made a far greater donations, the gift of his time, experience, and expertise.

"I get great satisfaction out of interacting with students . . . really with anyone with an interest in entrepreneurship," said Mark. "To earn those opportunities, I told Duncan I was committed to making sure the program would grow."

"It was really a partnership that made this happen," said Duncan. "I could drive the academic part, but Mark was driving funding and other issues, particularly with the Board of Trustees and the president."

The effort was aided when, in 2007, Mark was asked to join the University of Rochester Board of Trustees. So, while his major impact could be seen within the Entrepreneur Program, it is also seen within the university as a whole.

"Through that whole process things really started moving," said Duncan. "We were able to hire additional staff people using some of his money to pay for them initially. As a result, we were able to secure contracts and grants . . . a total of about four million dollars, which is a great return on investment."

"Today," said Mark, "Rochester is the kind of place where students from all of the schools, not just business and engineering, have the opportunity to learn about entrepreneurship. It extends opportunities to students in the realms of education, medicine, and many other arenas. Because coming up with an idea isn't just about making money. You need to make money, too. But some ideas can change the way people live. And through our partnership we're able to give a much broader spectrum of people a chance to make their own difference."

Take, for instance, the story of a company now formed as Health Care Originals. The founders of this company met as graduate students at Rochester in 2013. Though they initially didn't have any idea how to create a business plan, they did discover an innovative idea for a wearable asthma monitoring solution within the University of Rochester Patent Office. As enrollees in the MS Technical Entrepreneurship and Management program, they learned how to create a business plan around their concept. Little did they know that being named one of that's year's three winners in the Ain Business Plan Competition would come with an extra prize . . . an investor.

"They didn't know that I had severe asthma, or how it had affected my life," recalled Mark. "But they won the competition and, when they incorporated, I invested in their company because I knew that my parents would have paid anything to have been able to have this technology available for me when I was young."

"By changing the culture of the university, Mark changed the trajectory of hundreds, maybe more, of individuals who gained a different

way of thinking . . . a different way of approaching things these days,"
said Duncan. "And maybe, because of this, these students have taken
slightly different paths than they would have taken otherwise. Again,
thanks to Mark, this program has shown students what you can do
when you graduate with a degree from Rochester beyond a traditional
path."

Soon, this incubator space for budding businesspeople was thriving. Students were being empowered to think both inside and outside
the box. As they did, they interacted with many people who likewise
heard, and marched, to the beat of a different drummer. And the
world was taking notice . . . and benefiting.

One such student was Daphne Pariser, who had made a trip
to Kenya as a ten-year-old. Struck by the poverty she had witnessed,
Daphne wanted to make a difference. Through her involvement in
the Ain Incubator, she was able to form Humans for Education, a nonprofit organization dedicated to helping those in need, providing educational opportunities, modernizing villages, and solving the societal
issues of poverty in Africa.

"She is just incredible," marveled Mark. "She has such passion.
Humans for Education is truly making a difference. And we didn't just
help her get started . . . we also make a donation every year!"

These relationships, and many others, were formed based on good
people with great ideas, and Mark's commitment to sharing his knowledge and experience in building companies with a next generation.
Through Rochester and beyond, his name can be found on several
board of director lists, places like SiMPore, Inc., a Rochester-based
company that likewise leveraged the incubator as it researched and
developed nano-technology that creates synthetic membranes.

"Some of the technology that's being developed is just a whole
new level," said Mark. "And those people, and their companies, need
the guidance to get up and running so that they can also change the
world."

In turn, the University of Rochester has benefited as its reputation
as a place that drives innovation and new ventures continues to rise.

"One of the other things that Mark's involvement has driven,"
lauded Duncan, "is that Rochester has moved into the Top Ten among
entrepreneurship program rankings according to the *Financial Times*
of London. Before he was involved, that was unthinkable!"

That doesn't mean that everyone who participates in the many
facets of the program must go off and start a business. In fact, according to Duncan, that's not the point. Rather, having an understanding

for how businesses are started ... detailing the many steps necessary to turn an idea into commerce ... is valuable information no matter where one's education leads. He cites an interesting statistic his department has gleaned from following up with students postgraduation. Among those who did start a business, a majority ended up starting a business based on something that was not in their Rochester business plan.

"What we've been able to accomplish is that we've been able to push entrepreneurship throughout the university. It's become part of the overall culture of the university. We started out with a large contingent within the faculty against the concept. But we've been able to successfully move the needle to the point where now we get many faculty members coming to us in the Entrepreneur Center to talk about their business ideas.

"And now we've also been able, with Mark's help, to establish a program for members of our extended community ... nonstudents. This was initially funded by Mark for its first few years but we ended up with a grant from the federal government to promote and assist budding entrepreneurs in both rural communities and urban environments.

"And for those that take a deeper dive, the chance to interact with someone like Mark is equally valuable, if not more so."

"We talk a lot about the money, the support, and showing up at meetings," said Duncan, "but I believe Mark gets his greatest satisfaction out of meeting and interacting with students. So he not only supports the Entrepreneur Program, he also supports students. And I know he's interacted with many students after they have graduated from Rochester."

To bolster that deep level of interaction, Mark opted to expand his collegial involvement and now counts Tufts University in Massachusetts and Florida Gulf Coast University as places where he interacts with entrepreneur-based programs. And, according to Duncan, Mark is able to show many aspiring businesspeople the best way to put forth their ideas.

"What Mark brings to the table, along with his great grasp of building great teams and writing great business plans, is that you can break out on your own and be successful. What we're stressing," detailed Duncan, "is that entrepreneurship is a way of thinking."

Since stepping down from Kronos, that thinking has extended beyond classrooms for Mark as well. Forever the entrepreneur, he has mentored and served on the boards of many start-ups, everything from companies engaged in the latest cutting-edge technology to the formation of the high-end limo company that was subsequently merged with another limo company. And a chance meeting at a neighborhood party with a recent college grad led to the formation of a real estate holding company that now boasts fourteen buildings and over 100 apartments in areas north and west of Boston.

Mark enjoyed that venture so much that he then replicated the effort in Florida, where he started a real estate development company with a builder in southwest Florida at the bottom of the 2008 economic downturn. And that real estate company has also done very well.

Mark points to another random interaction that again drew him into yet another venture.

"About 12 years ago," he recounted, "I met a recent graduate of Rochester who had managed to get his undergraduate and MBA degrees in just four years . . . a remarkable achievement. I assisted in helping him find a job at a venture capital firm specializing in assisting small businesses, mostly family-owned operations that wanted to create liquidity. Over time, his firm morphed into a micro-private equity company. They capitalized and raised money for each business. And this formula worked for many small businesses around the country. So, through him and his organization, I personally invested in, and served as a board member and mentor to, many of these small companies, again because I genuinely liked the interaction and the chance to help out.

"I get a lot of satisfaction helping budding entrepreneurs think through ideas as, together, we figure out how they can start a company," said Mark. "Mind you, I'm not looking for a job! So, I limit my time. I'll spend time evaluating someone's business proposal, or on calls talking with students about how Kronos got started, listening to ideas, and giving advice. When these people become successful and I know I've played a role, I feel successful too."

Medical students, music majors, future teachers, engineers, and business people . . . each with a potential game-changing idea . . . or two . . . or three . . . or more! Some ideas, properly cultivated, could provide the spark that will result in a success story.

Chapter Twenty-Two

WORDS FROM THE HEART

A career, like life itself, is a journey. But sometimes even the fast track has its speed limits.

—Mark Ain

By 2005, Kronos was a worldwide leader in its field. And its founder and CEO Mark Ain was recognized in business circles as a bona fide visionary and success story. So, it was only natural that the University of Rochester should tap him as the commencement speaker that year at its MBA graduation ceremony. He had, after all, not only achieved success on his own terms—he had also been a member of the school's first-ever MBA graduation class.

His words that day were directed at an audience that was witnessing its own technological revolution. Very few people built their own computers from scratch because they were much easier to purchase, and at a price point where virtually everyone had one. The internet had already redefined almost all aspects of everyday life. And new visions—YouTube, for example, was launching its first video—were being realized at a pace that only a decade earlier would have been considered unimaginable, and even crazy.

Against that backdrop, Mark took the stage and looked out over a crowd of starry-eyed graduates and their proud family members and friends and entertained and challenged them in words that could be applied to graduates in virtually any year.

The following is the text of that speech:

Thank you . . . Wow . . . what an honor . . . what a pleasure . . .
   To be back here at the Simon School.
   Back here at Rochester.
   And *not* have a paper due or an examination looming . . . just like you!
(laughter and applause)
   Feels pretty good, doesn't it? And as a Simon School alum myself, I know all too well that it wasn't easy.

I see a few bags under the eyes out there. As a former Simon School King-of-the-All-Nighters, I can relate to that as well!

You've walked the path that few can follow, and through your efforts, you've set yourselves apart from the pack—Simon School MBAs—poised to change this increasingly competitive and international world of business like few others.

Before me, I see a sense of pride and accomplishment. Enjoy it . . . and never forget it . . . because you deserve it . . .

I jumped at the chance to speak with you today—as I jumped at the chance to join the Simon School's Executive Advisory Committee a few months ago—because we all, individually *and* collectively, sit at a crossroads . . . and because all of us . . . together . . . hold so much potential for positive change.

But let me issue a quick disclaimer: when you first heard that I would be your speaker, you no doubt conjured up an image of a very tall, exceedingly handsome, polished, suave and dynamic speaker. But I must make a confession—I'm not really that tall.

No, the truth be told, while I regularly speak to customers and investors, this is my first-ever commencement speech. And because I want to get it right, I've written my remarks out. So, if you catch me reading somewhere along the way, or if I have a stumble or two, please, just ask yourself, "Exactly how tall is he?"

I ran into a quote recently that seemed appropriate to share with you today: "Affluent college-bound students face the real prospect of downward mobility. Feelings of entitlement clash with the awareness of imminent scarcity. There is resentment at growing up at the end of an era of plenty coupled with reassessment of conventional measures of success."

Needless to say, you are not "college bound." But can anyone relate?

Because this quote, written by a man named Jimmy Holzer, was written in 1950. In other words, the more things change, the more they stay the same. And one man's seemingly bleak outlook is probably someone else's opportunity.

As you heard, I'm the founder and CEO of Kronos. If you will, allow me to give you a quick overview: Kronos, named for the Greek god of time, came into being because *way back* in 1977 we identified a need in the workforce management arena. *Time clocks.* The only aspect of paying employees that had not been automated at the time. And so, we began experimenting, and eventually came up with a clock that could record and calculate the time and attendance of what was then called "frontline labor"—basically the nonprofessional end of the workforce—those who actually punched in and out of work on a daily basis . . .

But you must know, ours was not an overnight success story. The original clocks did collect time. But they also occasionally caught fire.

As you might imagine, this presented a slight hitch in our early marketing plans.

Our first office and "manufacturing facility" was an old ironworks plant—really a *glorified garage*—behind the Harvard Business School in Boston. And I still remember the day we received our first order: a small copy center in New York City, on Broadway, no less, hung that baby right at the back of the store in a place of prominence, just above the toilet. *We couldn't have been more proud!*

Of course, we'd dreamed of that day and had prepared by hiring my brother Aron—today the COO and Executive Vice President of Kronos——because we needed an *official Kronos delivery vehicle* . . . and Aron *owned a hatchback!*

But, from those humble—and oftentimes humbling—beginnings, we have never lost sight of our need to consistently take strides to improve. With technology as the key driving force in our operations and goals, we put out the fires and successfully reinvented the scope and processes of our operation, and in so doing, continuously improved the value proposition offered to our client partners.

And because we never saw the status quo as good enough . . . because we used our sense of pride and accomplishment as motivation to break what wasn't broken and improve at every possible opportunity, we eventually made the leap from the outhouse to the penthouse!

From one clock resting over a toilet to a point where Kronos now assists tens of thousands of companies worldwide "improve the performance of their people and their businesses" by better managing and optimizing their workforces, Kronos is today used by over 20 million people around the globe—20 million people!

Needless to say, we're now much more than time clocks. By taking measured steps, and deploying technology at the right time, Kronos today offers a fully integrated suite of applications that cover the entire life cycle of the employee, from hiring to scheduling and tracking, from analyzing productivity to ensuring payroll accuracy, Kronos is there, and adding value, every step of the way.

Today, we have over 2,700 employees . . . 67 offices across the country and around the world . . . and recently we were able to very proudly report our 101st consecutive quarter of quarter-on-quarter, year-on-year revenue growth.

Making us one of only four publicly held software companies to have grown year-to-year and been profitable for the past twelve years or more, with a track record for revenue and profitability growth that's exceeded by only one other technology company: a little outfit called Microsoft.

Now that's Kronos. But the question here today is: What lesson in my story relates to you and where your journey may one day take you?

To get there, I'm going to tell you a quick story about my family, particularly my grandmother, because she was so influential on my family in general, and on me in particular.

Many, many years ago, my teenaged grandfather emigrated from what is now Eastern Russia in advance of his bride-to-be . . . my grandmother. Once he felt like he'd established himself—in New York City—he sent for my grandmother, who was eighteen years old at the time. As soon as she arrived, she asked what type of work he was doing. He told her not to worry, that he was working hard for a good company, and that they would be okay and able to pay the bills.

Well, that may have been good enough for some people. But my grandmother wasn't "some people."

She said that's not enough. We're not getting married until you own your own business. Imagine that?! She was eighteen!

Well, my grandfather wasn't going to lose a woman like my grandmother. And so, with her as his partner in life *and* business, the two of them established their own business, making and hanging slip covers and drapes. And they poured their hearts and souls into making themselves successful . . . eventually saving enough of their profits to buy not only their own place, but several apartment buildings as well.

And the entrepreneurial spirit of the Ain family was successfully brought to this country.

My father ended up running a plumbing supply business that my grandfather founded . . . while my mother graduated high school at fourteen years old . . . college at seventeen . . . and became one of the first woman to graduate from the Columbia School of Law at twenty . . . and passed the New York State bar exam before her twenty-first birthday!

Along their journey, which I'm happy to say has brought them to live nearby in Massachusetts recently, they raised five children, of which I'm the oldest, and instilled us with these basic principles that I tried to instill in Kronos . . .

To be passionate in everything we do . . .

To be driven to execute the task before us, while constantly focusing on the road ahead . . .

To be honest, with ourselves, our customers, and our investors . . .

To be apolitical within our corporate culture, always focusing on what's best for the business . . .

And to be conservative in our approach, understanding the many potential ramifications of the decisions we make.

Those characteristics have shaped the success story that is Kronos. And even though Kronos has grown to be a key player in a worldwide marketplace, I'd like to think it's still the kind of company that my grandmother would have been proud of . . . *even if she wouldn't have allowed my grandfather to work for us.*

My point here is that there's an entrepreneur in each of us. And while it takes a leap of faith to follow that dream, it also requires the right preparation, and the right help.

So, a first piece of advice . . . don't be anxious.

A career, like life itself, is a journey. But sometimes even the fast track has its speed limits.

I didn't leave the Simon School and found Kronos. I worked at the former Esso International Company, then at the former Digital Equipment Company, in jobs I really didn't like, but in positions that afforded me the opportunity to work with and observe some excellent people.

After those stints, I went into consulting, first joining a company, then starting my own firm. Work, but also more golden opportunities to study the "real" world.

Even what would be considered my "leap" was less a leap than a carefully measured step . . . the result of three long years of screening product ideas before discovering . . . finally . . . that time—and the time clock—were both on my side.

Kronos became the result.

You'll rarely get to use all you've learned here at the Simon School in your first few years in the business world. And it would be easy to get frustrated by that. But don't.

View those first few jobs as the continuance of your education . . . as hands-on training.

Work with the best possible people at the best possible organization. Seek out a mentor or role model and follow their lead.

Is that a sacrifice . . . to commit to a few experiences that you might not necessarily love? You bet it is. But tempering the urge to leap too soon will ultimately benefit you in the long run.

Because even though it's said that "it's not how you start, it's how you finish," that's not necessarily true. Because how and where you finish will almost always be affected by how you start. So, start smart . . .

That said, today's business climate is different, and changing in ever-decreasing cycles.

Without question, technology will affect your individual trajectories. Which means you'd better be on your toes. Because technological shifts are like natural disasters . . . you can look ahead a little . . . and prognosticate a lot. But ultimately, it's next to impossible to accurately predict that next big leap.

So, you have to be nimble . . . and prepared for change . . .

For example, who could have predicted microprocessors? We didn't. But at Kronos we immediately made them integral in the design of our data collection terminals . . . and our little company was suddenly on a course toward $100 million in annual revenue.

The proliferation of PCs in both the workplace and the home? Again, at Kronos, an opportunity . . . and soon we were able to centralize the data those terminals were collecting—and our revenues surged, again.

Then along came something called the internet! And while others struggled with what to do with this "information superhighway," we jumped into the fast lane, reinventing our solutions to reach the professional end of the employee base, and offer our customers a solution for their entire workforce. The results were amazing!

So, look ahead with five- and ten-year plans . . . but keep your sharpest eye on the near term. Because suffice it to say, the fact that there's no internet-like revolution in focus in the short term doesn't mean that technology won't revolutionize the world tomorrow . . . or the next day.

We'll be doing the same thing at Kronos. Ready to fix what isn't broken. Because once a technology-based company has hung its one and only product over a toilet in New York . . . praying that it won't catch on fire . . . you know enough not to focus too closely on long-range plans.

Now the Greek God of Time would probably say that I've held up this celebration far too long . . . that I've delivered a pile of advice on a day that's better suited for celebrating than lecturing. But please let me impose on you with one more very important notion:

In everything you do, in every decision, in every interaction, in every step . . .

Be ethical.

Sounds simple enough. But it isn't. The higher you rise, the higher the stakes, the greater the challenges . . .

Be ethical.

Thanks to your parents and those others who have supported you through the years, and thanks to the great educators here at the Simon School, each of you will most likely earn at least more than what 99 percent of the population here in the richest country in the world earns. You'll be executives—CEOs, CFOs, COOs, presidents—people of power. Don't abuse it.

As you rise to positions of leadership, understand that with authority comes responsibility . . . to your employees, your shareholders, and yourselves.

We have to get back to a time when people equated leadership with doing the right thing. So be opportunistic. But don't take shortcuts. Never turn away a paycheck, or a stock option. But be certain you've earned what you get.

Years ago, I attended a seminar that included a great exercise. We were told to sit down and create our own personal Board of Directors, made up of people—living or dead—whose opinions we would value most. Then, when making key decisions, the idea was to hold a brief mental meeting with your Board and listen to the advice they might give you.

Try it. It works, and it gives you great perspective. And if you like, feel free to borrow one name from my Board: Dora Ponemon. Because even though my grandmother had been long dead when Kronos was first formed, her entrepreneurial spirit was a guiding force for me and what became my Kronos. And I'm sure she would be proud to be associated with so many future leaders today.

In parting, I wish each and every one of you congratulations and all the success and happiness this world has to offer.

Now let's all make this a Simon School day to remember . . .

Thank you.

Chapter Twenty-Three

THE FOUNDATION

*May you have a strong foundation when the winds of change shift
. . . and may you be forever young.*

—Bob Dylan

No structure can be considered sound without a solid foundation. Good, firm foundations resist change, provide support, and create value and integrity. This, of course, is true in building. Whether for a lakeside cottage or a massive superstructure, the underpinnings on which anything is built are critical to the lasting value and appeal of every physical creation.

All that would seem self-evident, especially for those who rise to the top of their chosen fields, yet such is not always the case. Often, blind ambition obscures what is most basic, and therefore, most important.

Take Frank Lloyd Wright. Easily considered the greatest architect in American history, Wright's ambition to create a masterpiece homage to what would become known as organic architecture somehow shaded his view of what was the "right" thing to do when it came to his design of a home that would come to be known as Fallingwater.

At 67 years of age, Wright was already renowned for his designs when he was commissioned by business mogul Edgar J. Kaufmann Sr. to create a summer home on Kaufmann's rural Pennsylvania property. By that point, Wright had designed nearly a thousand homes. But this was to be special. The house would be constructed at the top of a waterfall.

Wright nearly withdrew from the project after Kaufmann brought in an engineering firm to check his plans. The firm suggested that the structure's foundation would not be strong enough. Wright, ever more concerned with the art of the project, vehemently disagreed. So, to his credit and without Wright's knowledge, Kaufmann had extra reinforcement added to the foundation. The resultant building was, to Wright's vision, a true masterpiece, documented by the Smithsonian on its 1991 "Life List of 28 Places to Visit before You Die." Yet, were it not for Kaufmann, the structure would likely not have survived.

What Wright lost sight of, Mark Ain did not. Mark understood the basic tenet that everything needs a solid foundation. "Everything" means that he used that axiom to build Kronos and his life. Yes, he spent an inordinate amount of time getting the company created, then securing funding, building an engineering team, then a sales force . . . but he did so all the while making hires that proved to stand the test of time, thus creating a solid corporate foundation built on domain knowledge, work ethic and dedication to constant improvement.

One such hire was his long-time administrative assistant Cheryl Ferruccio, whose close personal perspective on Mark covered much more than the business side of things.

"I've worked directly with and for Mark for over twenty-five years," said Cheryl. "In fact, I sometimes talk to Mark more than I talk with my husband!"

Cheryl said that prior to her interview with Mark way back when, she was "warned" that Mark was a "little different." But she quickly learned that Mark was an interesting combination of seemingly contradictory traits: quirky but decisive, demanding but understanding, driven but free-spirited.

"I learned very quickly that Mark was both easy to talk to and easy to listen to," said Cheryl. "He ran the company and created a culture within Kronos. He's just a normal, likeable person who cares for people. And that shows at Kronos. The company is such an ethical place where everyone works well together and everyone does what needs to be done to be successful. And all of that is a reflection of the person Mark is."

Cheryl also noted that Mark's style begat an early corporate trend.

"The Kronos culture of casual dress started with Mark," she divulged. "Back when everyone was still wearing suits and ties, he was himself. And he liked orange!"

To be sure, Mark could sometimes be found at headquarters sporting bright orange pants accented with either sandals or matching orange sneakers.

"He was so proud of those bright orange pants!" Cheryl exclaimed.

"Dressing down," said Mark, "actually became a selling point for Kronos, because it was casual. I'd ask people, 'Why are you getting dressed up for work?'"

What Cheryl admired about Mark more, though, was that he didn't simply have a foundation for his company.

"He was so proud of his family, and especially his children. He was very family oriented," she said.

Mark readily admits that, like most families, things weren't perfect. He and his first wife, Lillian, opted to divorce in 1991 but not until they had three children, two sons, and a daughter—Josh, Adam, and Rebecca. Travel time spent securing, financing, and building a sales network took its toll. But he took pride in making the time to be a basketball and soccer coach, and a dad, and to help the kids through some tough transitions.

"I liked that about him as much as any success he earned through Kronos," recalled Cheryl.

Seems all the Ain kids were headstrong and free spirited. Yet, each was able to find his or her path to success as they each defined it.

Today, each is in a great position. Josh and then Adam both attended top-ranked Williams College. Josh, who started out his career as a programmer, has a senior position at Google. Adam, after earning an MBA at Stanford, is the chief operating officer of a start-up in Washington, DC. Rebecca, who attended prestigious Vassar College, found fulfillment in teaching.

"All three kids," said Mark, "are doing just fine."

For his own part, post-split, Mark eventually decided to test the field.

"I put a relationship ad in *Boston Magazine* in the fall of 1992," Mark divulged. "But nothing worked out. So, in the fall of 1994, I decided to run another ad. And, because there were so many voice mails, I asked an assistant to listen to the responses and make her recommendations."

"He did," confirmed Cheryl. "And one stood out so I encouraged Mark to reach out to Carolyn. Then, a bit later, I helped plan the wedding!"

Carolyn, a former executive recruiter, quickly brought Mark balance . . . a new foundation for his life. Gregarious and engaging, her smile is infectious, particularly to Mark. They've been together for twenty-six years and enjoy children, grandchildren, traveling and staying home in equal measure.

"It's just great to see him happy," said Cheryl. "I love his family. They've always treated me like family. And I know they care for my family as well."

In retrospect, Cheryl feels as though that first, somewhat trepidatious, meeting was a turning point in her life, not because of the steady employment but much more so as a result of the long-term relationship.

"In addition to the loyalty I feel for Mark, one of his greatest attributes, one that I have appreciated more than anything else, has been his loyalty to me," she said. "I feel valuable to him and he includes me in pretty much all of his business matters, whether for Kronos, outside boards, or other outside business matters. This has helped me to become more efficient and has always made me feel part of his team. For that, I am extremely grateful."

"I've earned a lot of money . . . more than I can ever spend," Mark said humbly, "and I think that when you achieve a level of success like I have that it's vitally important to give back."

In Mark's case, that giving back came in many forms. He contributed time, expertise, and money to the University of Rochester. And, in 2015, the powers-that-be at Rochester honored Mark by rededicating the building housing the Entrepreneurship Program and the incubator space as the Ain Center for Entrepreneurship.

"I did not expect that!" Mark confessed. "It came as a real surprise when I received that phone call! And the event was fantastic, but not because they were honoring me. The best part was how many colleagues and friends were invited and managed to make it there that day. To be able to interact with all these people, some who'd helped me and many students that I had interacted with and assisted . . . it was just a great day!"

Mark subsequently contributed that same three-pronged support of time, knowledge, and financial support, albeit on a smaller scale, to the entrepreneurial programs at Tufts University and Florida Gulf Coast University.

In a different arena, after having emergency surgery at a Lee Memorial Health Systems hospital in Fort Meyers, Florida, Mark was so impressed by the care he received from the certified nursing assistants (CNAs) on staff that he contacted the Lee Memorial Health System Foundation and asked what he could do to honor and thank them.

"I was told that there had been a program planned to provide extra training for the CNAs. But, because their budget had been cut, they had to delay the implementation of that program until they could find a way to fund it," Mark recalled. "And I said, 'Well, you just did,' and now that program, which is designed to improve communication between CNAs and nurses as well as with patients and their families is making a difference in all five of their hospitals."

In the wake of that successful effort, Mark and Carolyn decided to sponsor a new aspect of Lee Memorial Health's annual employee recognition program and provided funding so that chosen CNAs would be granted scholarship money so they could enroll in nursing school and become RNs.

"A lot of very talented CNAs simply can't afford nursing school," reported Mark. "So, the awards ceremony is just great! When winners are announced all of them go crazy. I recall one woman in particular who'd worked her way up through several jobs. She never dreamed she'd be able to go to nursing school to become an RN. She was so happy when her name was called that she just burst into tears!"

Every year, Mark and Carolyn make it a point to attend the event.

"We go back to the awards ceremony and people who've gotten awards in past years come up and hug us," said Mark. "It's just a very uplifting experience."

So uplifting that the Ains decided to also offer a similar scholarship to CNAs at Massachusetts' Lahey Clinic, where Carolyn had served as a long-time trustee.

On top of all the start-ups and small businesses Mark has been involved in as a mentor, trustee, and investor, he has also made time to perform similar functions on the nonprofit side.

For many years he was a Trustee for the Walker School, a Massachusetts-based nonprofit that provides therapeutic and educational programs for children facing emotional, behavioral, and learning challenges, and likewise headed their investment committee.

He's served as both a Board Member and Chairman of the Investments Committee for the Boston and Waltham, Massachusetts, chapters of the YMCA.

And, with Carolyn, family members, and Cheryl as his rocks, it was only natural he and Carolyn would found the Ain Family Foundation.

"Over the years, as I helped him with some of his investments, he always used to tell me that 'Some day you will help me give away this money.' I never totally knew what he meant," she said, "but now I am honored to be a part of the Ain Family Foundation."

The charitable foundation has, through the years, made myriad donations to the arts, education, healthcare, and youth services. In fact, in a given year, the number of organizations that benefit from the Ain Family Foundation exceeds seventy-five. And though Mark won't disclose the full amount donated in a given year, it is substantial.

Some organizations like the Charles River Watershed Association, Friends of Middlesex Fells Reservation, the Massachusetts Audubon

Society, and the Nature Conservatory are focused on preserving the environment.

Others, like various chapters of Habitat for Humanity, the National Brain Tumor Society, the National Consumer Law Center, the Combined Jewish Philanthropies, Doctors Without Borders, the Foundation for Fighting Blindness, and many food banks, help those with physical ailments and limitations.

Still within education, Mark and Carolyn also contribute to the Aspen Institute, the Colorado-based think tank world-renowned for gathering thought leaders and leaders together to address a multitude of the world's key issues. And they support Columbia University in honor of his mother Pearl, who was one of the youngest as well as one of the first women to graduate from its School of Law.

And then there are the arts, as groups like the Florida Repertory Company, the Barrington Stage Company, the Urbanity Community Project, and the New Repertory Theatre also garner support.

"It's just very important to give back," said Mark. "And there are just so many worthwhile organizations. But I think if you were to go through the entire list, you'd see that quite a few of our causes provide programs that impact children and young adults because Carolyn and I share a passion for helping and educating young people."

To be sure, youth-oriented recipients including Horizons for Homeless Children, Inc., Cradles to Crayons, Big Brothers Big Sisters, Boston's Children's Museum, Teach for America, Outward Bound, and Squash Busters are also on the list of beneficiaries.

AFS-USA and AIESEC both promote studies abroad and international internships; the Bridge Charter School, Camp Seagull, and Camp Sunshine at Sebago Lake each assist young people in finding the right path; and Horizons for Homeless Children, Humans for Education, the Hyde Leadership Charter School, and the Foundation for the Developmentally Disabled are all integral in providing support where support is needed for young people.

"Both Carolyn and I really enjoy the chance to make a difference in the lives of others," said Mark. "We enjoy sharing the process with family and friends. And again, when you have achieved a level of success, it's something you should do."

"It is so satisfying seeing how many organizations are blessed with their donations," said Cheryl. "That's a big part of who they are as well. Mark and Carolyn have dedicated so much of their time to the foundation and to supporting so many deserving organizations. They have a very clear mission statement and do so much. It's just been incredible to play a part in all of it."

That the paths and outlets he has chosen later in life bring him somewhat full circle is not lost on Mark. As a boy, his asthma taught him to deal with adversity. His passion for reading and education in general provided him with a sense of self-worth. Along his own personal journey, he learned the value of support, and of mentors, and that education, hard work, and believing in someone other than yourself were all valuable elements in the recipe for success.

True, for Mark, like for everyone, the winds of change have shifted many times over the years. But through the journey, he enhanced and fortified his personal and professional foundations. He believes Kronos, in whatever form and moniker adopted as time moves on, is built to last. That makes him happy, and at least a little bit proud. And he's pleased with his life today . . . a successful life, firmly grounded by family and friends, in which each new day holds the possibility for teaching just one more person the difference between being a mule or a horse.

Because it is in the telling of his tale and the sharing of his adventures and insights that Mark too remains forever young.

EPILOGUE

Coming to the end of this project is extremely bittersweet. Through this process Mark and I have shared a journey, one that has endured an injury and health scare, an economic downturn, and even a pandemic.

It's been an honor and a privilege to have worked with Mark as he's reflected on his past—to gain a closer look at the factors and decisions that drove and influenced him. He's someone I've long admired because he did indeed break the mold to create something bigger than himself, a task very few attempt and far fewer achieve.

Deeper than that, and I think the tone of the conversations that took place between Mark and his core Kronites proves this out, is the underlying premise that you can come to relish hard work and long hours if you have a sense of accomplishment and camaraderie as key ingredients in whatever you set out to accomplish. Because these were good people. Talented? Yes. Diligent? You bet. But they were also fun-loving, decent folk ... the type you'd happily have a beer with, take on a fishing trip, invite over for dinner, or join in logging long hours to solve a complex business problem. And if you can mix all those factors into your professional and personal life, you're part of something that's truly special. And that is what Kronos was and, remarkably or not, still is.

Is there a lesson in here that can be replicated? Absolutely. Does this book represent an etched in stone blueprint for success? Not necessarily. Because times change. Values change. But within the context of these moments of reflection by Mark and all the participants, there is at least a series of basic building blocks from which success can be derived.

First, climb your own ladder of success. Don't blindly go hand over foot up someone else's ladder, because before you go to the effort, you must be sure the ladder is taking you where you want to go. And even after you've created your own rungs, and have begun the daily process of aspiring to go higher each day, you must occasionally step back to ensure that your version matches your vision and is, therefore, taking you to the right destination. Because individual priorities are seldom carved in granite. Everyone's definition of success mutates as life goes on. And if at the end you've reached the goal you set for yourself a month, a year or even a few decades back doesn't culminate in you feeling happy, then what was the point?

Mark Ain is a happy man. He's not perfect. But he's perfectly fine with that. He's prideful, but not boastful. He loves and is loved. Even as he's crossed into the back end of his seventies, that keeps a spring in his step. That's what people should envy. Not the wealth. Not the success. The smile. The contentment. The happiness!

With its recent merger with Ultimate Software, Kronos entered giant territory. Revenue is now measured in billions. The number of people using the systems each day is equal to the populations of some continents. If gathered together as a town, Kronos would rank in the top ten percent of U.S. cities, towns, and villages size-wise. You'd think, given Mark's strategy during Kronos' fledgling years, that the global Godzilla of what's now termed human capital management, the Goliath of managing the virtual life cycle of the employee, would be ripe for the taking by some young gun and his merry band. But Kronos has that in check by being huge but acting small and hungry, burning with the same fire that Mark lit and tended for decades.

As a result of the merger with Ultimate Software, the Kronos name will be dispatched into the sunset, phased out in favor of a new moniker, UKG . . . Ultimate Kronos Group. It embodies this latest super-permutation, but comes with a tinge of irony. After spending so much time toiling in relative obscurity, just as the Kronos name was becoming known and recognized around the globe, it will slip inside a three-letter monogram. But the company that Mark built was always structured to promote growth, and embrace change, and so it goes.

We didn't cover all the many changes throughout Kronos history. That would take volumes as the company responded to both obstacles and opportunities, reset boundaries and expectations, and redefined what it meant, and means, for client organizations to get the most out of their most valuable, and costly, resource in the form of their workforces. This was more a chance to reminisce . . . to get a sense for the personalities that meshed, sometimes seamlessly but often through coordinated effort, to blaze a path to sustained success.

At this point, Mark realizes that things have been successful beyond even his grandest of dreams. But rather than feel fulfilled, he feels challenged . . . challenged to help others in various ways. But playing to his strengths, he wants to be the spark to the entrepreneurial fires that burn within younger people.

Could a new wave of like-minded free thinkers exist out there that, given the right motivation and guidance, will likewise prosper into a new and dynamic organization capable of accomplishing what Mark and those early Kronites did? Absolutely! Will having read this tome provide some prerequisite motivation and guidance? Maybe. Truthfully, creating something from scratch and then riding it forward for decades will be different today.

But it can and will be done. New pioneers are born every day. With each sunrise someone tells themselves that they think a better mousetrap can be built, that their ideas have merit, even when everyone around them questions the logic behind bucking the norm. And the beauty is, as I hope this collection of remembrances prove, that dreams of success can be shared, and realized collaboratively, meaning from one idea many trajectories can be changed for the better.

As the outset of this journalistic journey, we detailed Mark's favorite quote from his favorite philosopher, the ancient Chinese philosopher and writer Lao-Tzu:

> A leader is best when people barely know he exists," the philosopher espoused, "not so good when people obey and acclaim him. Worse when they despise him. But of a good leader who talks little, when his work is done and his aim fulfilled, they will say, "We did it ourselves."

Words to live, and work, by? They were, if your name was Mark Ain.

So, if you do happen to bump into Mark in the not-so-distant future, whether at Costco or the local sandwich shop, and you recognize him from the pictures in this book, feel free to engage him. Because, without a doubt, he still has stories to tell . . . and inspiration to share.

Thanks!